T0234430

Mass Spectrometry–Based Glycoproteomics and Its Clinic Application

Mass Spectrometry–Based Glycoproteomics and Its Clinic Application

Edited by
Haojie Lu

CRC Press
Taylor & Francis Group
Boca Raton London New York

CRC Press is an imprint of the
Taylor & Francis Group, an **informa** business

First edition published 2022
by CRC Press
6000 Broken Sound Parkway NW, Suite 300, Boca Raton, FL 33487-2742

and by CRC Press
2 Park Square, Milton Park, Abingdon, Oxon, OX14 4RN

© 2022 selection and editorial matter, Haojie Lu; individual chapters, the contributors

CRC Press is an imprint of Taylor & Francis Group, LLC

Reasonable efforts have been made to publish reliable data and information, but the author and publisher cannot assume responsibility for the validity of all materials or the consequences of their use. The authors and publishers have attempted to trace the copyright holders of all material reproduced in this publication and apologize to copyright holders if permission to publish in this form has not been obtained. If any copyright material has not been acknowledged please write and let us know so we may rectify in any future reprint.

Except as permitted under U.S. Copyright Law, no part of this book may be reprinted, reproduced, transmitted, or utilized in any form by any electronic, mechanical, or other means, now known or hereafter invented, including photocopying, microfilming, and recording, or in any information storage or retrieval system, without written permission from the publishers.

For permission to photocopy or use material electronically from this work, access www.copyright.com or contact the Copyright Clearance Center, Inc. (CCC), 222 Rosewood Drive, Danvers, MA 01923, 978-750-8400. For works that are not available on CCC please contact mpkbookspermissions@tandf.co.uk

Trademark notice: Product or corporate names may be trademarks or registered trademarks and are used only for identification and explanation without intent to infringe.

Library of Congress Cataloging-in-Publication Data

Names: Lu, Haojie, 1974- editor.
Title: Mass spectrometry based glycoproteomics and its clinic application / edited by Haojie Lu.
Description: First edition. | Boca Raton, FL : CRC Press, 2021. | Includes bibliographical references and index. | Summary: "This book demonstrates the progress that has been achieved in Mass spectrometry based glycoproteomic analysis, including a wide range of analytical methodologies and strategies involved in selective enrichment, qualitative analysis, quantitative analysis and data analy.sis, together with their clinical applications. The book will serve as a valuable resource to aid appropriate techniques and their applications for students, postdocs, and researchers working in proteomics, glycoscience, analytical chemistry, biochemistry and clinical medicine"-- Provided by publisher.
Identifiers: LCCN 2021001208 (print) | LCCN 2021001209 (ebook) | ISBN 9781032028613 (hardcover) | ISBN 9781003185833 (ebook)
Subjects: LCSH: Glycoproteins--Analysis. | Mass spectrometry--Methodology.
Classification: LCC QP552.G59 M35 2021 (print) | LCC QP552.G59 (ebook) | DDC 572/.68--dc23
LC record available at https://lccn.loc.gov/2021001208
LC ebook record available at https://lccn.loc.gov/2021001209

ISBN: 978-1-032-02861-3 (hbk)
ISBN: 978-1-032-02923-8 (pbk)
ISBN: 978-1-003-18583-3 (ebk)

DOI: 10.1201/9781003185833

Typeset in Minion
by KnowledgeWorks Global Ltd.

Contents

Editor

Haojie Lu, PhD, is a Professor at Fudan University since 2003. He earned his PhD from Lanzhou Institute of Chemical Physics, Chinese Academy of Sciences in 2001. He specializes in proteomics based on mass spectrometry with particular emphasis on novel technologies for separation and identification of low-abundant proteins and post-translationally modified proteins (including glycosylation), as well as relative and absolute quantification methods for proteomics. Dr. Lu's awards include 2009 Winner of Young Chemist Award from China Chemical Society; 2009 University Distinguished Professor of Shanghai (Eastern Scholar); 2010 supported by The National Science Fund for Distinguished Young Scholars; and 2013 he was awarded the Eastern Scholar tracking program. He has authored articles published in national and international journals.

Contributors

Weiqian Cao
The Fifth People's Hospital, Fudan
  University, and The Shanghai
  Key Laboratory of Medical
  Epigenetics
The International Co-laboratory
  of Medical Epigenetics and
  Metabolism
Ministry of Science and Technology
Institutes of Biomedical Sciences
NHC Key Laboratory of
  Glycoconjugates Research
Fudan University
Shanghai, China

Caiyun Fang
Department of Chemistry
Fudan University
Shanghai, China

Haojie Lu
Department of Chemistry
Institutes of Biomedical Sciences
NHC Key Laboratory of
  Glycoconjugates Research, and
  Shanghai Cancer Center
Fudan University
Shanghai, China

Suideng Qin
School of Chemical Science
  & Engineering
Shanghai Key Laboratory of
  Chemical Assessment and
  Sustainability
Tongji University
Shanghai, China

Shifang Ren
NHC Key Laboratory of
  Glycoconjugates Research
Department of Biochemistry
  and Molecular Biology
School of Basic Medical Sciences
Fudan University
Shanghai, China

Zhenyu Sun
Institutes of Biomedical Sciences
Fudan University
Shanghai, China

Zhixin Tian
School of Chemical Science
  & Engineering
Shanghai Key Laboratory of
  Chemical Assessment and
  Sustainability
Tongji University
Shanghai, China

Guoli Wang
Institutes of Biomedical Sciences
Fudan University
Shanghai, China

Pengyuan Yang
Institutes of Biomedical Sciences
  and Department of Chemistry
Fudan University
Shanghai, China

Ying Zhang
Institutes of Biomedical Sciences,
  and NHC Key Laboratory of
  Glycoconjugates Research
Fudan University
Shanghai, China

Selective Enrichment Methods for N-Glycosite Containing Peptides in N-Glycoproteomics

Guoli Wang, Ying Zhang, and Haojie Lu

CONTENTS

1.1 INTRODUCTION

As one of the most important and ubiquitous post-translational modifications (PTMs) of proteins, glycosylation plays a critical role in various biological processes, including cell adhesion, recognition, receptor activation,

DOI: 10.1201/9781003185833-1

1

and so on (Ohtsubo and Marth 2006). Also, aberrant glycosylation has been closely associated with the development and progression of various diseases, such as cancers (Pinho and Reis 2015), neurodegenerative diseases (Hwang et al. 2010), and so on. It is estimated that more than 50% of the proteins are glycosylated (Apweiler et al. 1999), and almost all of the key molecules involved in the innate and adaptive immune response are glycoproteins (Marth and Grewal 2008; Rudd et al. 2001). In addition, many available Food and Drug Administration (FDA)-approved cancer biomarkers are glycosylated proteins, being of much clinical use to diagnose, monitor, and use in therapies (Ludwig and Weinstein 2005). Given all of the above, glycoproteome research is of great significance in the field of life science. Nowadays, mass spectrometry (MS) technique has been increasingly developed to analyze glycoproteins and has become the main tool in glycoproteomics research (Dell and Morris 2001; Lu et al. 2016). However, compared with non-glycoproteins, most glycoproteins usually present at a low abundance (Alvarez-Manilla et al. 2006). Furthermore, the wide dynamic range and heterogeneity of glycosylation make it a considerably challenging task to identify and quantify glycoproteins. Therefore, the enrichment of glycoproteins/glycopeptides from complex bio-samples is indispensable to achieve a comprehensive analysis of glycoproteome before MS analysis. A growing number of enrichment methods have been developed and exhibit great potential for global profiling of the N-glycoproteome. In this chapter, the current and representative development in enrichment strategies of MS-based glycoproteomics is reviewed and summarized. These methods can be roughly divided into covalent, non-covalent, and tagging-assisted enrichment, as shown in Figure 1.1.

1.2 NON-COVALENT ENRICHMENT

Non-covalent enrichment methods are based on non-covalent interaction between the glycoproteins/glycopeptides and polar materials or affinity materials. Polyhydroxy structure of glycans endows glycopeptides with high polarity and hydrophilicity and makes it possible to be enriched by solid support through the interactions like Van der Waals forces and hydrogen bonds. Benefiting from the reversible capture and release of glycopeptides, non-covalent enrichment methods can retain the intact structure of glycopeptides with glycan information and have been employed widely in glycoproteomics research. Two main non-covalent methods,

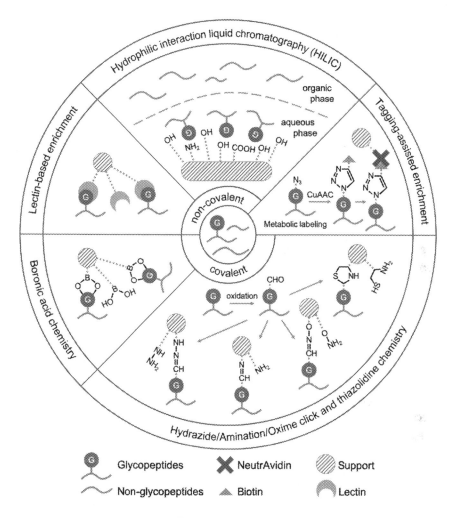

FIGURE 1.1 Separation and interaction modes for the enrichment of glycoproteome.

lectin-based enrichment and hydrophilic interaction liquid chromatography (HILIC), are discussed below.

1.2.1 Lectin-Based Enrichment

Lectins are carbohydrate-binding proteins mostly from plants and have been applied widely in glycoproteomics research due to the ability to recognize some specific glycans with a unique motif. Unlike an antibody targeting a particular protein, each lectin can orient towards a certain type of glycan on the glycopeptides or glycoproteins. Therefore, the selectively capture and release of glycopeptides can be performed using various functional materials modified with lectins, including magnetic nanoparticles (MNPs) (Dong et al. 2015), graphene oxide (GO) (Shi et al. 2015), and silica gel (Liu et al. 2015). A silica-based material (SiO₂-ODex Con A), with oxidized dextran as the spacer and concanavalin A as the affinity reagent, which binds to high-mannose-type and bi-antennary complex-type glycans, has been synthesized and evaluated using ovalbumin (OVA), demonstrating the promising potential of glycosylation analysis (Liu et al. 2016). However, only the use of single lectin to enrich glycoprotein/glycopeptides may usually cause biased enrichment. Hence, to attain comprehensive analysis, a combination of diversified types of lectins can be a better choice, which has been certified by a strategy based on FASP (filter aided sample preparation) and multi-lectin affinity (Figure 1.2).

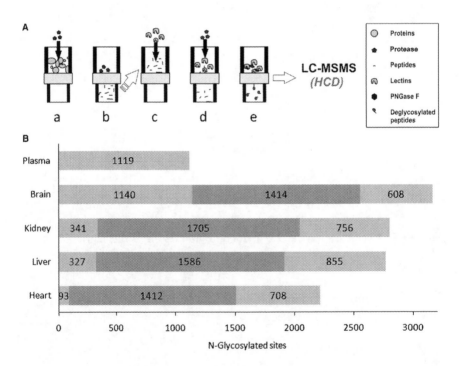

FIGURE 1.2   (A) Sample preparation and enrichment of N-glycosylated peptides: N-Glyco-FASP and (B) N-glycoproteome of different mouse tissues. The data was composed of three parts: sites only detected in this organ; sites detected in at least two organs; sites both detected in blood plasma (Zielinska et al. 2010).

As a consequence, a total of 6,367 N-glycosites on 2,352 proteins were successfully identified and mapped in four mouse tissues as well as plasma (Zielinska et al. 2010).

Similarly, three different types of lectins immobilized on magnetic nanoprobes (MNP@ConA, MNP@AAL, and MNP@SNA) were employed associatively for profiling N-glycoproteome of non-small cell lung cancer (NSCLC) (Waniwan et al. 2018), and the combination of Con A and wheat germ agglutinin (WGA) was utilized for glycoprotein enrichment from human serum (Yang and Hancock 2004). Other than these common lectins, some novel molecules can also recognize the glycan moieties on proteins specifically. As a member of the E3 ubiquitin ligase family, Fbs proteins preferably bind to denatured glycoproteins modified with high-mannose or complex-type glycans, participating in the ER-associated degradation (ERAD) pathway for the removal of unfolded N-glycoproteins (Yoshida et al. 2005). Recently, through mutagenesis and plasmid display selection, Fbs1 GYR variant, a mutant of Fbs1, showed higher affinity than wild-type Fbs1 for large-scale N-glycoproteomics studies, and a comparison against the above-mentioned lectin enrichment method from Matthias group was conducted using 1 μg IgG-depleted human serum by three MS runs. Ultimately, 2,559 unique intact N-glycopeptides were identified by Fbs1 GYR enrichment relative to 1,172 by lectin enrichment (Chen et al. 2017). Fbs1 GYR variant possessed higher affinity and improved recovery of complex N-glycoproteins, demonstrating its potential to be employed for substantially unbiased enrichment of N-linked glycopeptides from bio-samples.

1.2.2 HILIC-Based Enrichment

The hydrophilic glycan configuration equips glycopeptides with a better ability to interact with polar materials than non-glycosylated peptides. Therefore, separation and enrichment of glycopeptides from peptides mixture can be achieved by principle based on HILIC, with non-glycosylated peptides passed through while glycopeptides retained on the solid phase under appropriate condition (Selman et al. 2011). Compared with lectin affinity chromatography (LAC), HILIC generally displays broader specificity to glycan and gets access to a larger-scale glycoproteome. Various novel hydrophilic groups immobilized on different stationary materials have been developed recently, including metal organic frameworks (MOFs) (Hu et al. 2019; Wu et al. 2019; Zhou et al. 2020), covalent organic frameworks (COFs) (Ma et al. 2019; Wang et al. 2017), and magnetic mesoporous phenolic resin (MMP) (Zhang et al. 2018). It was reported that 418 N-glycopeptides from 125 glycoproteins were identified by magnetic Fe_3O_4 nanospheres coated with Mg-MOF-74 (Fe_3O_4@Mg-MOF-74) from 1 μL human serum, owing to the hydrophilicity from oxygen atoms and the open metal sites (Li et al. 2017). While COFs, composed of light elements by strong covalent bonds, are supposed to be more stable than MOFs, a magnetic COF (mCTpBD), prepared recently with abundant nanopores and rapid magnetic responsiveness, exhibits a high sensitivity of 0.5 fmol/μL for enriching glycopeptides from standard horseradish peroxidase (HRP) (Figure 1.3) (Wu et al. 2020b).

Supramolecular organic frameworks (SOFs) have also been applied for the enrichment of glycopeptides, ascribed to their high loading capacity, large surface area, and good biocompatibility. As reported newly, a bifunctional gallium ion immobilized magnetic SOF (magOTfP5SOF-Ga^{3+}) was synthesized by complexation reaction to simultaneously enrich glycosylated and phosphorylated peptides, whose abundant sulfonic acid groups

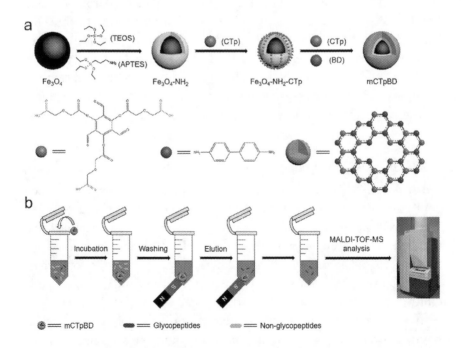

FIGURE 1.3   (A) Graphical synthetic route of mCTpBD; (B) enrichment procedure of glycopeptides by mCTpBD (Wu et al. 2020b).

and gallium ions can interact with glycans and phosphate, respectively (Zheng et al. 2020). Other than these organic frameworks materials with great size exclusion effect originating from the abundant porous structure (Wang et al. 2016), polymer is another alternative for HILIC materials selection. A branched copolymer with unique claw-like polyhydric groups, Sil@Poly(THMA-co-MBAAm), was prepared by thiol-ene click reaction, which captured 1,997 N-glycopeptides corresponding to 686 glycoproteins from the digests of the mouse brain (Shao et al. 2016). To promote the performance of polymers in glycopeptides enrichment, a polymeric monolithic tip modified with piperazine was synthesized through the free radical polymerization for the enrichment of low-abundance glycopeptides with a sensitivity of 100 attomoles (Sajid et al. 2020). Recently, zwitterionic hydrophilic interaction liquid chromatography (ZIC-HILIC), possessing additional charges on the surface compared to traditional HILIC, has been employed for glycoproteome research with outstanding performance (Li et al. 2019), of which iminodiacetic acid (Lin et al. 2018), glutathione (Liu et al. 2018), and cysteine (Guo et al. 2019; Wu et al. 2020a; Xia et al. 2018) have been reported a lot. For example, a novel zwitterionic hydrophilic material decorated with Fe_3O_4 and Au nanoparticles, followed by tripeptide L-glutathione (GSH) modification, has been designed for glycopeptides enrichment, with one-dimensional hydroxyapatite nanofiber (HN) as the sustainment, and 246 N-glycopeptides of 104 glycoproteins were identified by magHN/Au-GSH nanofiber from 1 µL human serum (Huan et al. 2019). Zwitterionic groups were also introduced to multiple branched poly(amidoamine) dendrimer (PAMAM), and 44 unique glycopeptides mapped to 28 glycoproteins were still detected from only 0.1 µL human serum, indicating the excellent capture capacity for glycosylation exploration from complex bio-samples can be achieved by the as-prepared ZICF-PAMAM (Cao et al. 2016). Combination of HILIC and other methods is considered available to amplify glycopeptides coverage, a strategy in which HILIC and electrostatic repulsion liquid chromatography (ERLIC) were connected in tandem, has been reported to identify and quantify the glycosylation for excavating potential cancer glycopeptide biomarkers (Zacharias et al. 2016). A lot of advantages, such as low biases to glycopeptides, high throughput, and good reproducibility, have been achieved by HILIC strategies. Thereby various new materials have been developed continually. However, to reach practical applications for glycopeptides enrichment from high matrix complexity of clinical samples, its inherent flaws remain to be addressed yet, including no discrimination between O- and N-linked glycopeptides and nonspecific retention of non-glycopeptides containing hydrophilic residues or some other hydrophilic biomolecules.

1.3 COVALENT-BOND FORMATION-BASED ENRICHMENT

Covalent enrichment methods are based on covalent bonds formation between glycopeptides and functional materials, thus exhibiting higher specificity and selectivity for glycopeptides than non-covalent methods. Generally, functional materials are modified with specific groups, which can covalently interact with glycans on glycopeptides to achieve the enrichment. Among them, boronic acid chemistry-based methods are reversible and thereby can preserve the glycans intact. Other irreversible covalent methods, such as hydrazide chemistry, oxime click chemistry, and amination chemistry, can wreck original glycan structure causing loss of glycan information, thus suitable for N-glycosylation sites and deglycosylated peptides analysis. Several main covalent enrichment methods are reviewed below.

1.3.1 Boronic Acid Chemistry

Boronic acid chemistry is one of the most popular covalent enrichment methods, in which five- or six-membered cyclic esters are formed between boronic acid and cis-diol compounds under alkaline conditions, while these cyclic esters dissociate under acidic conditions, which makes it possible to seize and release glycopeptides/glycoproteins with a solid support containing boronic acid groups by modulating the pH simply (Sparbier et al. 2005). Therefore, compared to the biased defects of LAC and nonspecific retention of HILIC, boronic acid methods may have a better performance for the enrichment of glycopeptides/glycoproteins which contain the cis-diol structure, and diverse functional materials decorated with boronic acid groups have been developed to date, including mesoporous silica (Xu et al. 2009), metal organic frameworks (MOFs) (Li et al. 2018; Xie et al. 2018; Yang et al. 2017), copolymers (An et al. 2020; Jiang et al. 2017; Zhang et al. 2015a), and dendrimers (Wang et al. 2013). As the main mechanism is the same, we will display some representative examples according to the matrix materials and different application purposes. As an important type of functional material, mesoporous nanoparticles have unique applications in protein/peptide enrichment with their advantages,

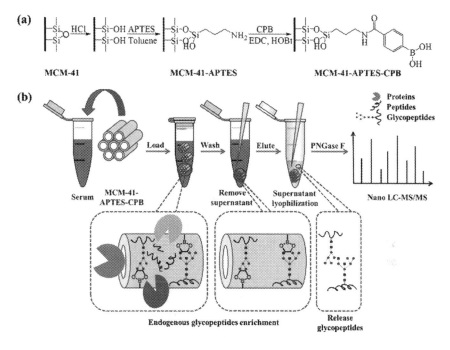

FIGURE 1.4 (A) Schematic overview of the preparation of boronic acid-functionalized MCM-41 mesoporous silica; (B) flowchart of the enrichment of glycopeptides followed by nano-LC-MS/MS detection (Liu et al. 2012).

including high surface area and narrow distribution of regular pore size. For glycopeptidome research, we synthesized a boronic acid grafted mesoporous nanoparticle using MCM-41 as the substrate (shown in Figure 1.4). With the combination of the size exclusion effect of MCM-41 and the selectivity of boronic acid chemistry, excellent selectivity (glycopeptide/non-glycopeptide analyses at molar ratios of 1:100) and good binding capacity (40 mg/g) have been achieved by this material (Liu et al. 2012).

To facilitate the separation of enriched glycopeptides, a temperature-responsive core cross-linked star (CCS) polymer with boronic acids in its soluble polymeric arms was reported for the enrichment of glycoproteins, and it is worth noting that this material exhibited rapid, reversible thermal-induced volume phase transition with temperature from 15 to 30°C, leading to high capture capacity of 210 mg/g within 20 minutes (Chen et al. 2019). To meet the demand for the enrichment of both phosphorylated and glycosylated peptides simultaneously, a novel magnetic MOF was designed and employed in complex biological samples, benefiting from the metal unit of Zr^{4+} ions as well as dual organic ligands of PTA and PBA (Luo et al. 2019). For highly efficient enrichment of glycopeptides and reducing the enrichment step, our group developed a new synergistic method combining two different nanomaterials, where boronic acid-functionalized Fe_3O_4 nanoparticles for enriching glycopeptides and poly(methyl methacrylate) nanobeads (PMMA) for strongly adsorbing non-glycopeptides, leading to extremely high sensitivity and selectivity for the capture of glycopeptides from the solution. Profiting from the separation of non-glycopeptides by PMMA nanobeads, the washing step was unnecessary, so that the enrichment process was simplified, and the recovery efficiency of glycopeptides reached 90% (Wang et al. 2014). General boronic chemistry-based enrichment methods need alkaline conditions, which are different from the neutral physiological conditions of glycoproteins, maybe resulting in some unpredictable degradation. For this reason, a novel CPBA-functionalized composite material (CPBA-Ni_6PW_9/SA) was developed, of which the abundant oxygen atoms significantly reduced the pKa value of CPBA, making it possible to perform isolation of original glycoproteins in a neutral medium with a maximum adsorption capacity of 495 mg/g (Xu et al. 2020). To improve the affinity of boronic acid toward cis-diol on glycopeptides, a comparison among several different boronic acid derivatives was conducted, revealing good performance in glycopeptides coverage of benzoboroxole. Subsequently, a dendrimer conjugated with benzoboroxole (DBA) was synthesized for synergistic interactions with glycopeptides in the enrichment process, and a total of 4,691 N-glycosylation sites on 1,906 glycoproteins were identified by combining the results from three human cell lines (MCF7, HEK 293T, and Jurkat) (Xiao et al. 2018). In addition to the enrichment of glycopeptides/glycoproteins on the omics scale, the selective enrichment of target glycoproteins have been achieved recently by employing available molecularly imprinted polymers (MIPs) (Wang et al. 2019b). A GO-based MIPs material (GO-APBA/MIPs) was designed and applied for OVA separation, where template glycoprotein OVA was pre-immobilized onto the surface of boronic acid-functionalized GO (GO-APBA) through boronate affinity followed by the deposition of imprinting layer through sol-gel polymerization, and after the removal of OVA, 3D cavities with double recognition abilities toward OVA and saturation binding capacity of 278 mg/g were achieved in the as-prepared imprinted materials

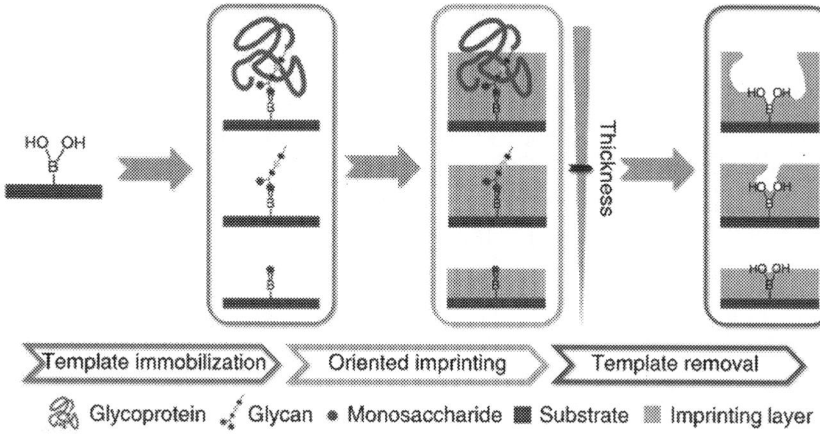

FIGURE 1.5   Schematic of boronate affinity controllable-oriented surface imprinting (Xing et al. 2017).

(Luo et al. 2017). To overcome the general uncontrollability of many imprinting approaches, an approach called boronate affinity controllable-oriented surface imprinting was developed (Figure 1.5), which allows for easy and efficient preparation of MIPs specific to glycoproteins, glycans, and monosaccharides. Meanwhile, with the controllable manner, the thickness of the imprinting layer could be fine-tuned according to the molecular size of the template by adjusting the imprinting time (Xing et al. 2017).

In summary, several prominent advantages have been obtained by boronic acid chemistry. First, many types of novel functional materials are available for boronic acid groups to conjugate with. Second, mild and controlled conditions make it easy to capture/release glycopeptides. Third, without enzymatically removing the glycans, the reversible reaction allows the released glycopeptides to directly provide specific glycans information at certain binding sites in MS analysis. However, boronic acid-functionalized materials often have associated nonspecific binding due to low binding constants and relatively low selectivity. Immobilization of boronic acid on low-specific adsorption surfaces would help to solve this problem.

1.3.2 Hydrazide Chemistry

Hydrazide chemistry-based methods are another important strategy for N-glycoproteome enrichment based on covalent interactions, which are thought to be more specific and unbiased than other enrichment methods due to high specificity and commercially available reagents. Compared with boronic acid chemistry, hydrazide chemistry is irreversible thus needs to cooperate with peptide-N-glycosidase F (PNGase F) generally to release enriched deglycosylated peptides for glycosites analysis. The first application of hydrazide chemistry in glycoproteins enrichment was developed in 2003 as shown in Figure 1.6, where cis-diol groups of glycans were firstly oxidized to aldehydes by periodate followed by hydrazone formation between aldehydes and hydrazide groups on a stationary support, then after on-bead digestion of immobilized proteins, non-glycopeptides were removed while enriched glycopeptides were liberated with PNGase F for MS identification and quantification (Zhang et al. 2003).

With the merit of hydrazide chemistry, various matrices have been exploited for highly selective enrichment of glycopeptides from complex

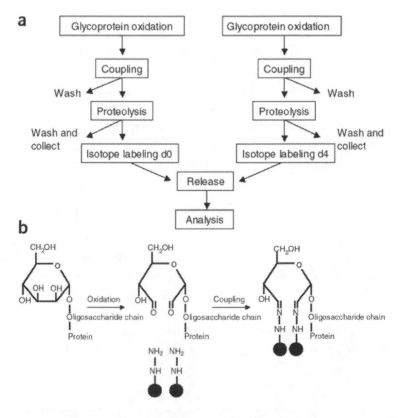

FIGURE 1.6 (A) Strategy for quantitative analysis of glycopeptides; (B) illustration of the hydrazide chemistry for glycoprotein enrichment (Zhang et al. 2003).

bio-samples. Our group introduced a novel hydrazide-functionalized core-shell magnetic nanocomposite, Fe_3O_4@poly(methacrylic hydrazide) (Fe_3O_4@PMAH), with polymer shell hanging abundant hydrazide groups, which began with a rapid and simple reflux precipitation polymerization to coat the magnetic core with carboxyl-terminal groups-contained polymer, followed by the generation of hydrazide groups through the reaction between surface carboxyls and adipic acid dihydrazides (ADHs). Taking advantage of high specificity from hydrazide-functionalized shell and convenient separation from the magnetic core, the as-prepared nanocomposites improved more than five times the signal-to-noise ratio of standard glycopeptides compared to commercially available hydrazide resin (Liu et al. 2014). Then, the same year, we designed another hydrazide-functionalized dendrimer material for efficient and selective enrichment of N-linked glycopeptides with FASP. The as-synthesized polyamido-amine (PAMAM) dendrimer materials possessed not only an ideal high molecular weight for thorough separation but also an appropriate molecular size which is comparable to many folded proteins, facilitating interactions during the incubation process, and 158 unique glycopeptides of 60 different glycoproteins from human serum had been mapped successfully by this hydrazide-functionalized PAMAM-based enrichment strategy (Zhang et al. 2014b). Other than these, different matrices, such as gold nanoparticles (Tran et al. 2012), graphene oxide (Bai et al. 2017), and polymers (Wang et al. 2019a), have also been reported to functionalized with hydrazide groups. Recently, for in-depth identification of N-glycosylation in human plasma exosome, a thermosensitive polymer with hydrazide groups was developed, which can completely dissolve to form a homogeneous solution for efficient interaction while precipitated easily by raising the system temperature to above 34°C, and 329 N-glycosites from 180 N-glycoproteins were identified from plasma exosomes, 26 glycoproteins of which were found changed significantly in glioma patients (Bai et al. 2018). Another pH-responsive soluble polymer-based homogeneous system was also reported for N-glycopeptides enrichment, and 1,317 N-glycopeptides from 458 glycoproteins in mouse brain were identified, more than twice by commercial solid/insoluble materials (Bai et al. 2015). In addition, galactose oxidase, which can specifically convert the hydroxyl group of C6 of Gal/GalNAc to an aldehyde group, is an ideal enzyme to tag glycoproteins in a mild condition. Instead of periodate, galactose oxidase has also been combined with hydrazide beads, with which global and site-specific analysis of cell surface N-glycoproteins was achieved, and 953 N-glycosites on 393 surface glycoproteins were identified successfully in MCF7 cells (Sun et al. 2019). For in-depth glycoproteome, hydrazide chemistry-based enrichment was combined with click maltose-HILIC to map the N-glycosylation sites of human livers, resulting in 14,480 N-glycopeptides with N-!P-[S/T/C] motif corresponding to 4,783 N-glycosylation sites on 2,210 N-glycoproteins identified (Zhu et al. 2014). Despite many advantages achieved, including high specificity and selectivity form covalent capture and easily combined with other enrichment strategies, like lectin, HILIC, and so on, hydrazide chemistry enrichment method requires a relatively long time for extra desalting step due to a large amount of salt added, and the irreversible reaction and enzymatic treatment to liberate peptides from a support (e.g., PNGase F) result in glycan information loss. In addition, possible side reactions during glycan oxidation from peptide backbones have been proved, such as aldehyde generation from amino alcohol of N-terminal Ser/Thr glycopeptides, which may have a serious impact on the performance of hydrazide chemistry, while N-terminal protection via dimethyl labelling may overcome this drawback (Huang et al. 2015).

1.3.3 Oxime Click Chemistry

Other than hydrazide groups, aminooxy groups also exhibit strong nucleophilicity due to the enhanced effect by the unshared pair on the adjacent atom (the α-effect) (Edwards and Pearson 1962), based on which classical oxime click reaction is carried out between an aldehyde group and an aminooxy group to produce oxime bond under mild conditions (Pauloehrl et al. 2012), and study has also shown that the product by oxime click chemistry is more stable than hydrazide chemistry (Kalia and Raines 2008). Its application in protein glycomics has been reported, in which the reductive terminal of glycan prior was labelled by aminooxy-contained biotinylated tag (Nishimura et al. 2005). Inspired from the glycoblotting method, an enrichment method focused on sialic acid (SA) terminated N-glycopeptides was developed by glycan-specific oxidation of SA residues of glycopeptides, followed by chemical ligation with aminooxy-functionalized polyacrylamide (Kurogochi et al. 2007). The oxime ligation could be dramatically accelerated with the presence of aniline, leading to efficiently labelling. And with the help of biotin-streptavidin enrichment, periodate-oxidized SA-containing glycans were selectively labelled and

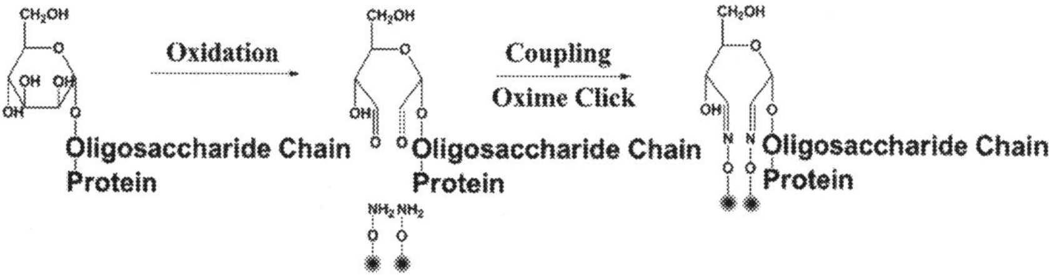

FIGURE 1.7   Illustration of the oxime click reaction for glycoprotein enrichment (Zhang et al. 2014a).

separated on the surface of living animal cells, while high cell viability remained due to the mild reaction condition at neutral pH (Zeng et al. 2009). By combining the periodate oxidation for SA and galactose oxidase for terminal galactose and N-acetylgalactosamine, an aniline-catalyzed oxime click reaction was employed to introduce aminooxy-functionalized biotin onto glycoproteins followed by streptavidin-coated beads capture, and a total of 175 unique N-glycosylation sites corresponding to 108 non-redundant glycoproteins were mapped with PNGase F treatment and MS identification (Ramya et al. 2013). Our group further improved this oxime click chemistry-based enrichment strategy by introducing a novel solid-phase extraction method (Figure 1.7), in which highly selective extraction of oxidized glycan chains on glycopeptides was achieved by oxime click reaction with newly synthesized aminooxy-functionalized magnetic nanoparticles, leading to excellent enrichment performance with high sensitivity of fmol level and high selectivity (extracting glycopeptides from mixtures of nonglycopeptides at a 1:100 molar ratio) while it took only 1 hour for coupling compared to 12–16 hours by hydrazide chemistry (Zhang et al. 2014a).

1.3.4 Amination Chemistry

Similar to hydrazide chemistry and oxime click chemistry, reductive amination chemistry-based method is another common strategy developed for glycan derivatization, in which an unstable Schiff base is formed firstly between glycan containing aldehyde group and primary amine under acidic condition, then reduced to a secondary amine with a reducing agent (AbdelMagid et al. 1996). For N-glycopeptides enrichment, our group introduced a solid-phase extraction strategy based on the reductive amination reaction between amino groups-functionalized magnetic nanoparticles and glycan whose cis-diol groups were oxidized in advance. Briefly, the amino groups from 3-aminopropyltriethoxysilane (APTES), easily assembled on the surface of nanoparticles by a one-step silanization reaction, could subsequently conjugate with oxidized glycopeptides to complete the extraction followed by treatment with PNGase F and MS analysis. In particular, the desalting step was eliminated in this protocol, therefore the extraction time was shortened to 4 hours compared with traditional hydrazide chemistry-based methods while improving the detection limit of glycopeptides by two orders of magnitude, and a total of 111 N-glycosylation sites from 108 glycopeptides were identified in 5 μL of human serum (Zhang et al. 2013). Similarly, our group then introduced APTES onto mesoporous silica to form a new functional material, SBA-15, for glycoproteins enrichment which was successfully employed to profile the N-glycoproteome of human colorectal cancer serum, while the coupling time was shortened from 4 to 1 hour with improved efficiency due to the original confinement effect of mesoporous material (Miao et al. 2016). However, a major limitation of this reductive amination reaction is the requirement for a proper reducing agent to form a stable C-N bond. As distinct from alkylamines, aniline can react with aldehyde group to generate a stable aromatic Schiff base in a wide pH range without reducing the double bond between carbon and nitrogen (Kaneshiro et al. 2011). We introduced a novel nonreductive amination-based method into N-glycoproteomics for the first time based on the reaction between aniline groups and oxidized glycopeptides, with a new type of aniline-functionalized nanoparticle (Zhang et al. 2015b).

1.3.5 Other Covalent-Chemistry

The majority of current methods for glycopeptides/glycoproteins enrichment have been reviewed above from several subheadings. And other than these, some strategies based on other principles have been reported recently. Our group developed a novel covalent enrichment method based on thiazolidine formation for ultrafast and efficient solid-phase extraction of the N-glycoproteome. Profiting from the facilely synthesized Cys-terminated magnetic nanoparticles, the glycopeptides could be selectively captured via the thiazolidine formation between the β-amino thiols groups on the surface of such nanoparticles and aldehyde groups on pre-oxidized glycan moieties of glycopeptides. Less than half an hour

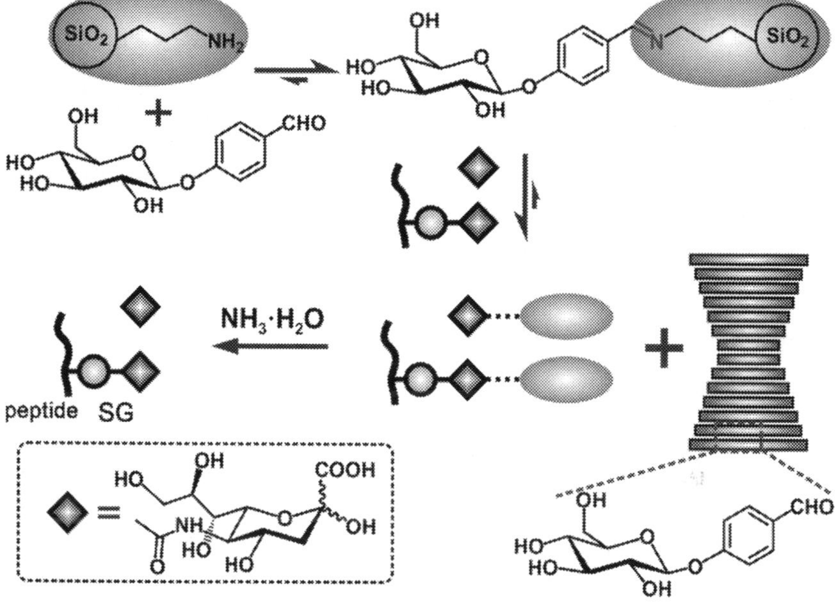

FIGURE 1.8 Schematic illustration of capturing strategy for sialic acid or SGs based on the Schiff base hydrolysis reaction (Xiong et al. 2020).

of coupling time was needed under mild conditions with no toxic catalyst or sample-destroying reducing agent added (Cai et al. 2020). To meet the urgent need for innovative and effective methods for precise capture of sialylated glycopeptides (SGPs) from bio-samples, a novel dynamic covalent chemistry strategy based on Schiff base hydrolysis was reported (Figure 1.8).

Firstly, the glucopyranoside-Schiff base-modified silica gel material (Glu-Schiffbase@SiO_2) was prepared via the formation of a reversible imine bond whose hydrolysis facilitated the formation of a unique sialylated glycan-precursor amine complex for SGPs enrichment. It is worth noting that the ordered self-assembly of hydrolysates and SG-precursor amine complex stabilized by hydrogen bond as well as electrostatic force promoted the hydrolysis reaction, thus achieving high selectivity (1:5000 with interference), high adsorption capacity (120 mg/g), and satisfying recovery of 95.5% with about 300 glycosites identified from 100 μg of human serum-extracted proteins (Xiong et al. 2020). However, it still remains unanswered whether the hydrolysis process is the prerequisite for the formation of the unique complex.

1.4 TAGGING-ASSISTED ENRICHMENT

By introducing a particular functional group into the glycan chain, the enrichment handle can be specifically labelled on glycopeptides via bioorthogonal chemistry. This type of enrichment strategy has been mainly used for cell surface glycoproteins research by metabolic labelling with commonly used azidosugar, such as N-azidoacetylgalactosamine (GalNAz) (Hubbard et al. 2011). Integrated with fast and specific click chemistry, like copper-free or copper-catalyzed azide-alkyne cycloaddition (CuAAC), tagged glycoproteins can be readily modified with biotin for NeutrAvidin enrichment, which has the potential for global and site-specific identification of cell surface N-glycoproteome (Smeekens et al. 2015). Such a combination of metabolic labelling and click chemistry as well as MS was also employed for systematic and quantitative analysis of the surface N-sialoglycoproteome in cancer cells with distinctive invasiveness (Chen et al. 2015). And recently, global characterization of surface glycoproteins from eight types of human cells has been achieved by this method, and a total of 2,172 N-glycosylation sites from 1,047 surface glycoproteins were identified successfully for correlation investigation between motif and protein secondary structures (Suttapitugsakul et al. 2019). Benefiting from this tagging-assisted enrichment and the natural abundance of the stable isotopes of bromine ($Br^{79}:Br^{81}=1:1$), an isotopic affinity probe has been prepared for intact glycopeptides analysis (Figure 1.9), consisting of four main parts: the alkyne group for reacting with azide on labelled glycopeptides via CuAAC, the biotin for avidin enrichment, the cleavable silane scaffold for the release of glycopeptides after enrichment, and the dibromide motif for detection in MS using a pattern-searching algorithm (Woo et al. 2015).

It is worth noting that different sugar analogs, including GalNAz, ManNAz, and GlcNAz, have distinct label ability, and evaluation showed that GalNAz displayed better performance in metabolic labelling compared with the other two (Xiao et al. 2016). In addition to proteomic identification, metabolic labelling and click chemistry can also be employed for glycoproteins imaging with a fluorescent group (Spiciarich et al. 2017) and systematic analysis of surface glycoproteins dynamics and half-lives with multiplexed proteomics (Xiao and Wu 2017).

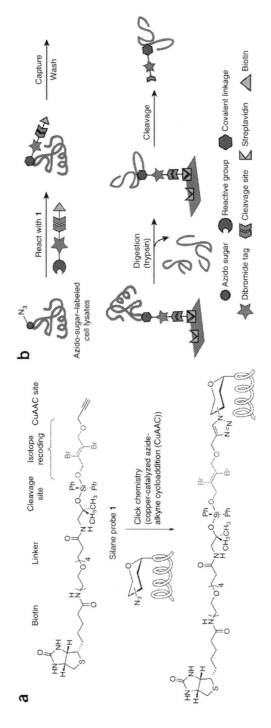

FIGURE 1.9 (A) Structure of cleavable probe 1 for reacting with azidosugar on glycopeptides via CuAAC; (B) enrichment strategy using dibrominated silane probe 1, shown in cartoon form (Woo et al. 2015).

1.5 CONCLUSIONS

The prevalence of glycosylation on a majority of proteins has been a research hotspot all the time, while large-scale analysis of N-glycoproteomics is always hard, with the challenge lying in its wide dynamic range of abundance and heterogeneity of various glycan structure. Based on the rapid advancement in chemical biology and MS technologies, increasingly comprehensive and in-depth analysis of glycoproteome has been achieved with a variety of enrichment strategies emerging. In this chapter, we have presented the mainstream enrichment methods for glycoproteins/glycopeptides including non-covalent methods based on affinity and hydrophilicity of glycans as well as covalent methods based on the chemistry of cis-diols on glycans. Besides, current progress was also reviewed with their advantages and limitations discussed, such as the superior specificity of lectin-based methods but bias enrichment, the high coverage by HILIC but nonspecific retention, and the good selectivity obtained by irreversible covalent methods but glycan information lost. In addition, great performance was also achieved by some unique methods. With the combination of recombinant glycosyltransferases (ST6Gal1) for labelling, and a correspondingly biotinylated sugar nucleotide for enrichment, a one-step selective exo-enzymatic labelling (SEEL) method was designed and employed for plasma membrane glycoproteins identification with exceptional efficiency and sensitivity (Sun et al. 2016). With these advances, MS-based glycoproteomics has provided plentiful valuable information from glycosylation sites to site-specific glycan structures. It is believed that advancement in methodologies will constantly promote the development of glycoproteomics and accelerate our exploration and thorough understanding of glycoproteins.

REFERENCES

AbdelMagid, A. F., K. G. Carson, B. D. Harris, C. A. Maryanoff, R. D. Shah. 1996. Reductive amination of aldehydes and ketones with sodium triacetoxyborohydride. Studies on direct and indirect reductive amination procedures. *Journal of Organic Chemistry* 61(11): 3849–62.

Alvarez-Manilla, G., J. Atwood, 3rd, Y. Guo, N. L. Warren, R. Orlando, M. Pierce. 2006. Tools for glycoproteomic analysis: Size exclusion chromatography facilitates identification of tryptic glycopeptides with N-linked glycosylation sites. *Journal of Proteome Research* 5(3): 701–8.

An, X., H. Wu, Y. Li, X. He, L. Chen, Y. Zhang. 2020. The hydrophilic boronic acid-poly(ethylene glycol) methyl ether methacrylate copolymer brushes functionalized magnetic carbon nanotubes for the selective enrichment of glycoproteins. *Talanta*, 210: 120632.

Apweiler, R., H. Hermjakob, N. Sharon. 1999. On the frequency of protein glycosylation, as deduced from analysis of the SWISS-PROT database. *Biochimica Et Biophysica Acta-General Subjects* 1473(1): 4–8.

Bai, H. H., C. Fan, W. J. Zhang, et al. 2015. A pH-responsive soluble polymer-based homogeneous system for fast and highly efficient N-glycoprotein/glycopeptide enrichment and identification by mass spectrometry. *Chemical Science* 6(7): 4234–41.

Bai, H. H., Y. T. Pan, C. Guo, et al. 2017. Synthesis of hydrazide-functionalized hydrophilic polymer hybrid graphene oxide for highly efficient N-glycopeptide enrichment and identification by mass spectrometry. *Talanta* 171: 124–31.

Bai, H. H., Y. T. Pan, L. Qi, et al. 2018. Development of a hydrazide-functionalized thermosensitive polymer based homogeneous system for highly efficient N-glycoprotein/glycopeptide enrichment from human plasma exosome. *Talanta* 186: 513–20.

Cai, Y., Y. Zhang, W. J. Yuan, J. Yao, G. Q. Yan, H. J. Lu. 2020. A thiazolidine formation-based approach for ultrafast and highly efficient solid-phase extraction of N-glycoproteome. *Analytica Chimica Acta*: 1100: 174–81.

Cao, W. Q., J. M. Huang, B. Y. Jiang, X. Gao, P. Y. Yang. 2016. Highly selective enrichment of glycopeptides based on zwitterionically functionalized soluble nanopolymers. *Scientific Reports*: 6: 29776.

Chen, W. X., J. M. Smeekens, R. H. Wu. 2015. Systematic and site-specific analysis of N-sialoglycosylated proteins on the cell surface by integrating click chemistry and MS-based proteomics. *Chemical Science* 6(8): 4681–9.

Chen, M., X. Shi, R. M. Duke, et al. 2017. An engineered high affinity Fbs1 carbohydrate binding protein for selective capture of N-glycans and N-glycopeptides. *Nature Communications* 8: 15487.

Chen, Y., J. Tong, J. Dong, J. Luo, X. Liu. 2019. A temperature-responsive boronate core cross-linked star (CCS) polymer for fast and highly efficient enrichment of glycoproteins. *Small* 15(13): e1900099.

Dell, A., H. R. Morris. 2001. Glycoprotein structure determination by mass spectrometry. *Science* 291(5512): 2351–6.

Dong, L., S. Feng, S. Li, P. Song, J. Wang. 2015. Preparation of concanavalin A-chelating magnetic nanoparticles for selective enrichment of glycoproteins. *Analytical Chemistry* 87(13): 6849–53.

Edwards, J. O., R. G. Pearson. 1962. The factors determining nucleophilic reactivities. *Journal of the American Chemical Society* 84(1):16–24.

Guo, P. F., X. M. Wang, M. M. Wang, T. Yang, M. L. Chen, J. H. Wang. 2019. Two-dimensional titanate-based zwitterionic hydrophilic sorbent for the selective adsorption of glycoproteins. *Analytica Chimica Acta*, 1088: 72–8.

Hu, X. F., Q. J. Liu, Y. L. Wu, Z. Q. Deng, J. Long, C. H. Deng. 2019. Magnetic metal-organic frameworks containing abundant carboxylic groups for highly effective enrichment of glycopeptides in breast cancer serum. *Talanta*, 204: 446–54.

Huan, W., J. Zhang, H. Qin, et al. 2019. A magnetic nanofiber-based zwitterionic hydrophilic material for the selective capture and identification of glycopeptides. *Nanoscale* 11(22): 10952–60.

Huang, J. F., H. Q. Qin, Z. Sun, et al. 2015. A peptide N-terminal protection strategy for comprehensive glycoproteome analysis using hydrazide chemistry based method. *Scientific Reports* 5: 10164.

Hubbard, S. C., M. Boyce, C. T. McVaugh, D. M. Peehl, C. R. Bertozzi. 2011. Cell surface glycoproteomic analysis of prostate cancer-derived PC-3 Cells. *Bioorganic & Medicinal Chemistry Letters* 21(17): 4945–50.

Hwang, H., J. Zhang, K. A. Chung, et al. 2010. Glycoproteomics in neurodegenerative diseases. *Mass Spectrometry Reviews* 29(1): 79–125.

Jiang, L., M. E. Messing, L. Ye. 2017. Temperature and pH dual-responsive core-brush nanocomposite for enrichment of glycoproteins. *Acs Applied Materials & Interfaces* 9(10): 8985–95.

Kalia, J., R. T. Raines. 2008. Hydrolytic stability of hydrazones and oximes. *Angewandte Chemie-International Edition* 47(39): 7523–6.

Kaneshiro, K., Y. Fukuyama, S. Iwamoto, S. Sekiya, K. Tanaka. 2011. Highly sensitive MALDI analyses of glycans by a new aminoquinoline-labeling method using 3-aminoquinoline/alpha-cyano-4-hydroxycinnamic acid liquid matrix. *Analytical Chemistry* 83(10): 3663–7.

Kurogochi, M., M. Amano, M. Fumoto, A. Takimoto, H. Kondo, S. Nishimura. 2007. Reverse glycoblotting allows rapid-enrichment glycoproteomics of biopharmaceuticals and disease-related biomarkers. *Angewandte Chemie-International Edition* 46(46): 8808–13.

Li, J., J. Wang, Y. Ling, et al. 2017. Unprecedented highly efficient capture of glycopeptides by Fe_3O_4@Mg-MOF-74 core-shell nanoparticles. *Chemical Communications* 53(28): 4018–21.

Li, S., Y. Qin, G. Zhong, C. Cai, X. Chen, C. Chen. 2018. Highly efficient separation of glycoprotein by dual-functional magnetic metal-organic framework with hydrophilicity and boronic acid affinity. *Acs Applied Materials & Interfaces* 10(33): 27612–20.

Li, X. W., H. Y. Zhang, N. Zhang, S. J. Ma, J. J. Ou, M. L. Ye. 2019. One-step preparation of zwitterionic-rich hydrophilic hydrothermal carbonaceous materials for enrichment of N-glycopeptides. *Acs Sustainable Chemistry & Engineering* 7(13): 11511–20.

Lin, H., X. Shao, Y. Lu, C. Deng. 2018. Preparation of iminodiacetic acid functionalized silica capillary trap column for on-column selective enrichment of N-linked glycopeptides. *Talanta* 188: 499–506.

Liu, L., Y. Zhang, L. Zhang, et al. 2012. Highly specific revelation of rat serum glycopeptidome by boronic acid-functionalized mesoporous silica. *Analytica Chimica Acta* 753: 64–72.

Liu, L. T., M. Yu, Y. Zhang, C. C. Wang, H. J. Lu. 2014. Hydrazide functionalized core-shell magnetic nanocomposites for highly specific enrichment of N-glycopeptides. *Acs Applied Materials & Interfaces* 6(10): 7823–32.

Liu, Y. J., D. M. Fu, Y. S. Xiao, Z. M. Guo, L. Yu, X. M. Liang. 2015. Synthesis and evaluation of a silica-bonded concanavalin A material for lectin affinity enrichment of N-linked glycoproteins and glycopeptides. *Analytical Methods* 7(1): 25–28.

Liu, Y., D. Fu, L. Yu, Y. Xiao, X. Peng, X. Liang. 2016. Oxidized dextran facilitated synthesis of a silica-based concanavalin A material for lectin affinity enrichment of glycoproteins/glycopeptides. *Journal of Chromatography A*, 1455, 147–55.

Liu, Q., C. H. Deng, N. Sun. 2018. Hydrophilic tripeptide-functionalized magnetic metal-organic frameworks for the highly efficient enrichment of N-linked glycopeptides. *Nanoscale* 10(25): 12149–55.

Lu, H. J., Y. Zhang, P. Y. Yang. 2016. Advancements in mass spectrometry-based glycoproteomics and glycomics. *National Science Review* 3(3): 345–64.

Ludwig, J. A., J. N. Weinstein. 2005. Biomarkers in cancer staging, prognosis and treatment selection. *Nature Reviews Cancer* 5(11): 845–56.

Luo, J., J. Huang, J. J. Cong, W. Wei, X. Y. Liu. 2017. Double recognition and selective extraction of glycoprotein based on the molecular imprinted graphene oxide and boronate affinity. *Acs Applied Materials & Interfaces* 9(8): 7735–44.

Luo, B., Q. Chen, J. He, et al. 2019. Boronic acid-functionalized magnetic metal-organic frameworks via a dual-ligand strategy for highly efficient enrichment of phosphopeptides and glycopeptides. *Acs Sustainable Chemistry & Engineering* 7(6): 6043–52.

Ma, Y. F., L. J. Wang, Y. L. Zhou, X. X. Zhang. 2019. A facilely synthesized glutathione-functionalized silver nanoparticle-grafted covalent organic framework for rapid and highly efficient enrichment of N-linked glycopeptides. *Nanoscale* 11(12): 5526–34.

Marth, J. D., P. K. Grewal. 2008. Mammalian glycosylation in immunity. *Nature Reviews Immunology* 8(11): 874–87.

Miao, W. L., C. Zhang, Y. Cai, Y. Zhang, H. J. Lu. 2016. Fast solid-phase extraction of N-Linked glycopeptides by amine-functionalized mesoporous silica nanoparticles. *Analyst* 141(8): 2435–40.

Nishimura, S. I., K. Niikura, M. Kurogochi, et al. 2005. High-throughput protein glycomics: Combined use of chemoselective glycoblotting and MALDI-TOF/TOF mass spectrometry. *Angewandte Chemie-International Edition* 44(1): 91–6.

Ohtsubo, K., J. D. Marth. 2006. Glycosylation in cellular mechanisms of health and disease. *Cell* 126(5): 855–67.

Pauloehrl, T., G. Delaittre, M. Bruns, et al. 2012. (Bio)Molecular surface patterning by phototriggered oxime ligation. *Angewandte Chemie-International Edition* 51(36): 9181–4.

Pinho, S. S., C. A. Reis. 2015. Glycosylation in cancer: Mechanisms and clinical implications. *Nature Reviews Cancer* 15(9): 540–55.

Ramya, T. N. C., E. Weerapana, B. F. Cravatt, J. C. Paulson. 2013. Glycoproteomics enabled by tagging sialic acid- or galactose-terminated glycans. *Glycobiology* 23(2): 211–21.

Rudd, P. M., T. Elliott, P. Cresswell, I. A. Wilson, R. A. Dwek. 2001. Glycosylation and the immune system. *Science* 291(5512): 2370–6.

Sajid, M. S., B. Jovcevski, T. L. Pukala, F. Jabeen, M. Najam-Ul-Haq. 2020. Fabrication of piperazine functionalized polymeric monolithic tip for rapid enrichment of glycopeptides/glycans. *Analytical Chemistry* 92(1): 683–9.

Selman, M. H., M. Hemayatkar, A. M. Deelder, M. Wuhrer. 2011. Cotton HILIC SPE microtips for microscale purification and enrichment of glycans and glycopeptides. *Analytical Chemistry* 83(7): 2492–9.

Shao, W. Y., J. X. Liu, K. G. Yang, et al. 2016. Hydrogen-bond interaction assisted branched copolymer HILIC material for separation and N-glycopeptides enrichment. *Talanta* 158: 361–7.

Shi, Z. M., C. Fan, J. J. Huang, et al. 2015. Preparation of graphene oxide based immobilized lectin and its application to efficient glycoprotein/glycopeptide enrichment. *Chinese Journal of Chromatography* 33(2): 116–22.

Smeekens, J. M., W. X. Chen, R. H. Wu. 2015. Mass spectrometric analysis of the cell surface N-glycoproteome by combining metabolic labeling and click chemistry. *Journal of the American Society for Mass Spectrometry* 26(4): 604–14.

Sparbier, K., S. Koch, I. Kessler, T. Wenzel, M. Kostrzewa. 2005. Selective isolation of glycoproteins and glycopeptides for MALDI-TOF MS detection supported by magnetic particles. *Journal of Biomolecular Techniques* 16(4): 407–13.

Spiciarich, D. R., R. Nolley, S. L. Maund, et al. 2017. Bioorthogonal labeling of human prostate cancer tissue slice cultures for glycoproteomics. *Angewandte Chemie-International Edition* 56(31): 8992–7.

Sun, T. T., S. H. Yu, P. Zhao, et al. 2016. One-step selective exoenzymatic labeling (SEEL) strategy for the biotinylation and identification of glycoproteins of living cells. *Journal of the American Chemical Society* 138(36): 11575–82.

Sun, F. X., S. Suttapitugsakul, R. H. Wu. 2019. Enzymatic tagging of glycoproteins on the cell surface for their global and site-specific analysis with mass spectrometry. *Analytical Chemistry* 91(6): 4195–203.

Suttapitugsakul, S., L. D. Ulmer, C. D. Jiang, F. X. Sun, R. H. Wu. 2019. Surface glycoproteomic analysis reveals that both unique and differential expression of surface glycoproteins determine the cell type. *Analytical Chemistry* 91(10): 6934–42.

Tran, T. H., S. Park, H. Lee, et al. 2012. Ultrasmall gold nanoparticles for highly specific isolation/enrichment of N-linked glycosylated peptides. *Analyst* 137(4): 991–8.

Wang, H. Y., Z. J. Bie, C. C. Lu, Z. Liu. 2013. Magnetic nanoparticles with dendrimer-assisted boronate avidity for the selective enrichment of trace glycoproteins. *Chemical Science* 4(11): 4298–303.

Wang, Y. L., M. B. Liu, L. Q. Xie, C. Y. Fang, H. M. Xiong, H. J. Lu. 2014. Highly efficient enrichment method for glycopeptide analyses: Using specific and nonspecific nanoparticles synergistically. *Analytical Chemistry* 86(4): 2057–64.

Wang, J., J. Li, Y. Wang, M. Gao, X. Zhang, P. Yang. 2016. Development of versatile metal-organic framework functionalized magnetic graphene core-shell biocomposite for highly specific recognition of glycopeptides. *Acs Applied Materials & Interfaces* 8(41): 27482–9.

Wang, H., F. Jiao, F. Gao, et al. 2017. Facile synthesis of magnetic covalent organic frameworks for the hydrophilic enrichment of N-glycopeptides. *Journal of Materials Chemistry B* 5(22): 4052–9.

Wang, M. Z., J. J. Gao, B. Zhao, S. Thayumanavan, R. W. Vachet. 2019a. Efficient enrichment of glycopeptides by supramolecular nanoassemblies that use proximity-assisted covalent binding. *Analyst* 144(21): 6321–6.

Wang, P., H. J. Zhu, J. X. Liu, et al. 2019b. Double affinity integrated MIPs nanoparticles for specific separation of glycoproteins: A combination of synergistic multiple bindings and imprinting effect. *Chemical Engineering Journal* 358: 143–52.

Waniwan, J. T., Y. J. Chen, R. Capangpangan, S. H. Weng, Y. J. Chen. 2018. Glycoproteomic alterations in drug-resistant nonsmall cell lung cancer cells revealed by lectin magnetic nanoprobe-based mass spectrometry. *Journal of Proteome Research* 17(11): 3761–73.

Woo, C. M., A. T. Iavarone, D. R. Spiciarich, K. K. Palaniappan, C. R. Bertozzi. 2015. Isotope-targeted glycoproteomics (IsoTaG): A mass-independent platform for intact N- and O-glycopeptide discovery and analysis. *Nature Methods* 12(6): 561–7.

Wu, Y. L., Q. J. Liu, C. H. Deng. 2019. L-cysteine-modified metal-organic frameworks as multifunctional probes for efficient identification of N-linked glycopeptides and phosphopeptides in human crystalline lens. *Analytica Chimica Acta*, 1061: 110–21.

Wu, Y., H. Lin, Z. Xu, Y. Li, Z. Chen, C. Deng 2020a. Preparation of zwitterionic cysteine-modified silica microsphere capillary packed columns for the on-column enrichment and analysis of glycopeptides in human saliva. *Analytica Chimica Acta*, 1096: 1–8.

Wu, Y., N. Sun, C. Deng. 2020b. Construction of magnetic covalent organic frameworks with inherent hydrophilicity for efficiently enriching endogenous glycopeptides in human saliva. *Acs Applied Materials & Interfaces* 12(8): 9814–23.

Xia, C., F. Jiao, F. Gao, et al. 2018. Two-dimensional MoS2-based zwitterionic hydrophilic interaction liquid chromatography material for the specific enrichment of glycopeptides. *Analytical Chemistry* 90(11): 6651–9.

Xiao, H. P., G. X. Tang, R. H. Wu. 2016. Site-specific quantification of surface N-glycoproteins in statin-treated liver cells. *Analytical Chemistry* 88(6): 3324–32.

Xiao, H. P., R. H. Wu. 2017. Quantitative investigation of human cell surface N-Glycoprotein dynamics. *Chemical Science* 8(1): 268–77.

Xiao, H., W. Chen, J. M. Smeekens, R. Wu. 2018. An enrichment method based on synergistic and reversible covalent interactions for large-scale analysis of glycoproteins. *Nature Communications* 9(1): 1692.

Xie, Y., Q. Liu, Y. Li, C. Deng. 2018. Core-shell structured magnetic metal-organic framework composites for highly selective detection of N-glycopeptides based on boronic acid affinity chromatography. *Journal of Chromatography A*, 1540: 87–93.

Xing, R. R., S. S. Wang, Z. J. Bie, H. He, Z. Liu. 2017. Preparation of molecularly imprinted polymers specific to glycoproteins, glycans and monosaccharides via boronate affinity controllable-oriented surface imprinting. *Nature Protocols* 12(5): 964–87.

Xiong, Y. T., X. L. Li, M. M. Li, et al. 2020. What is hidden behind Schiff base hydrolysis? dynamic covalent chemistry for the precise capture of sialylated glycans. *Journal of the American Chemical Society* 142(16): 7627–37.

Xu, Y. W., Z. X. Wu, L. J. Zhang, et al. 2009. Highly specific enrichment of glycopeptides using boronic acid-functionalized mesoporous silica. *Analytical Chemistry* 81(1): 503–8.

Xu, W., J. F. Cao, Y. Y. Zhang, Y. Shu, J. H. Wang. 2020. Boronic acid modified polyoxometalate-alginate hybrid for the isolation of glycoproteins at neutral environment. *Talanta*, 210: 120620.

Yang, Z. P., W. S. Hancock. 2004. Approach to the comprehensive analysis of glycoproteins isolated from human serum using a multi-lectin affinity column. *Journal of Chromatography A* 1053(1–2): 79–88.

Yang, Q., Y. Zhu, B. Luo, F. Lan, Y. Wu, Z. Gu. 2017. pH-responsive magnetic metal-organic framework nanocomposites for selective capture and release of glycoproteins. *Nanoscale* 9(2): 527–32.

Yoshida, Y., E. Adachi, K. Fukiya, K. Iwai, K. Tanaka. 2005. Glycoprotein-specific ubiquitin ligases recognize N-Glycans in unfolded substrates. *Embo Reports* 6(3): 239–44.

Zacharias, L. G., A. K. Hartmann, E. Song, et al. 2016. HILIC and ERLIC enrichment of glycopeptides derived from breast and brain cancer cells. *Journal of Proteome Research* 15(10): 3624–34.

Zeng, Y., T. N. C. Ramya, A. Dirksen, P. E. Dawson, J. C. Paulson. 2009. High-efficiency labeling of sialylated glycoproteins on living cells. *Nature Methods* 6(3): 207–9.

Zhang, H., X. J. Li, D. B. Martin, R. Aebersold. 2003. Identification and quantification of N-linked glycoproteins using hydrazide chemistry, stable isotope labeling and mass spectrometry. *Nature Biotechnology* 21(6): 660–6.

Zhang, Y., M. Kuang, L. J. Zhang, P. Y. Yang, H. J. Lu. 2013. An accessible protocol for solid-phase extraction of N-linked glycopeptides through reductive amination by amine-functionalized magnetic nanoparticles. *Analytical Chemistry* 85(11): 5535–41.

Zhang, Y., M. Yu, C. Zhang, et al. 2014a. Highly selective and ultra-fast solid-phase extraction of N-Glycoproteome by oxime click chemistry using aminooxy-functionalized magnetic nanoparticles. *Analytical Chemistry* 86(15): 7920–4.

Zhang, L. J., H. C. Jiang, J. Yao, et al. 2014b. Highly specific enrichment of N-Linked glycopeptides based on hydrazide functionalized soluble nanopolymers. *Chemical Communications* 50(8): 1027–9.

Zhang, X., J. Wang, X. He, L. Chen, Y. Zhang. 2015a. Tailor-made boronic acid functionalized magnetic nanoparticles with a tunable polymer shell-assisted for the selective enrichment of glycoproteins/glycopeptides. *Acs Applied Materials & Interfaces* 7(44): 24576–84.

Zhang, Y., M. Yu, C. Zhang, et al. 2015b. Highly specific enrichment of N-glycoproteome through a nonreductive amination reaction using $Fe_3O_4@SiO_2$-aniline nanoparticles. *Chemical Communications* 51(27): 5982–5.

Zhang, Q., Y. Huang, B. Jiang, et al. 2018. In situ synthesis of magnetic mesoporous phenolic resin for the selective enrichment of glycopeptides. *Analytical Chemistry* 90(12): 7357–63.

Zheng, H., J. Jia, Z. Li, Q. Jia. 2020. Bifunctional magnetic supramolecular-organic framework: A nanoprobe for simultaneous enrichment of glycosylated and phosphorylated peptides. *Analytical Chemistry* 92(3): 2680–9.

Zhou, Y. Y., Y. Xu, C. C. Zhang, A. Emmer, H. Q. Zheng. 2020. Amino acid-functionalized two-dimensional hollow cobalt sulfide nanoleaves for the highly selective enrichment of N-linked glycopeptides. *Analytical Chemistry* 92(2): 2151–8.

Zhu, J., Z. Sun, K. Cheng, et al. 2014. Comprehensive mapping of protein N-glycosylation in human liver by combining hydrophilic interaction chromatography and hydrazide chemistry. *Journal of Proteome Research* 13(3): 1713–21.

Zielinska, D. F., F. Gnad, J. R. Wisniewski, M. Mann. 2010. Precision mapping of an in vivo N-glycoproteome reveals rigid topological and sequence constraints. *Cell* 141(5): 897–907.

CHAPTER 2

Quantitative Methods for N-Glycosite Containing Peptides in N-Glycoproteomics

Zhenyu Sun, Ying Zhang, and Haojie Lu

CONTENTS

2.1 INTRODUCTION

Glycosylation plays vital roles in cells, including the determination of protein folding, trafficking, and stability, as well as regulation of nearly every extracellular activity such as cell-cell communication and cell-matrix interactions (Pinho and Reis 2015). Aberrant protein glycosylation, including the abundance of glycoprotein, degree of glycosylation, and structure of glycans, is closely related to various diseases, such as cancer,

DOI: 10.1201/9781003185833-2

neurodegenerative disorders, pulmonary diseases, blood disorders, and genetic diseases (Engle et al. 2019; Stadlmann et al. 2017; Stowell et al. 2015; Varki 2017). To provide deep insights into biological processes across the kingdoms of life, it is vital to study the glycoproteome quantitatively. Mass spectrometry (MS) plays a central role in glycoproteomic analyses. Developments in MS-based technologies have considerably advanced the quantification of glycoprotein in a high-throughput manner (Lu et al. 2016; Ruhaak et al. 2018).

The quantitative analysis of glycosylation usually faces the following challenges. First, glycopeptides typically have relatively lower abundance and lower ionization efficiency compared with non-glycopeptides. In complex biological samples, detection and quantification of these glycoproteins are usually hindered by many highly abundant non-glycoproteins. Therefore, highly efficient enrichment strategies are required to isolate glycoproteins/glycopeptides before MS analysis, as described in Chapter 1. Second, glycosylation is a non-template-driven process, which generates enormous diversity in their structures. The attached glycan further complicates the glycoproteome, as each glycosylation site on a specific glycoprotein contains various glycans. As a result, the microheterogeneity of glycans also raised the technical difficulty for analysis. Therefore, a strategy specific to glycoproteomics should be tailored towards its quantitative analysis.

Quantitative analysis of protein glycosylation can be performed at different levels for different purposes. Based on different quantitative purposes, quantitative methods can be roughly divided into two categories, i.e., relative quantification and absolute quantification. Relative quantification determines the change in the expression of glycoproteins between different samples by directly comparing the ion abundances between different samples. Absolute quantification measures the absolute amount of a specific glycoprotein in the sample. Further, according to different labelling methods, these quantitative methods could be divided into label-based and label-free approaches (Figure 2.1). These methods have been extensively used in glycoproteomics. To be noted, we only discuss how to quantitatively analyze the deglycosylated glycopeptides in this chapter, including both relative and absolute quantification. The quantitative analysis of intact glycopeptides is discussed in Chapter 4.

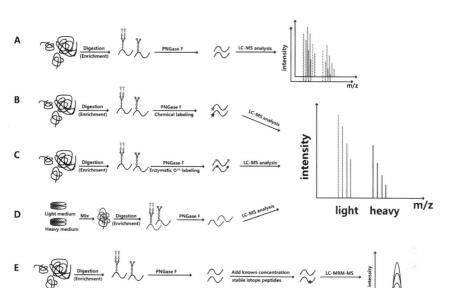

FIGURE 2.1 Overview of the different quantification methods. (A) Workflow of the label-free quantification method. (B) Workflow of the chemical labelling method. (C) Workflow of enzymatic O^{18}-labelling method. (D) Workflow of metabolic labelling method. (E) Workflow of the targeted quantification method. (A–D) are relative quantification methods and (E) is usually combined with the addition of reference glycopeptides into samples and used for absolute quantification.

2.2 RELATIVE QUANTIFICATION

In relative quantification, we analyze changes in protein expression in given samples relative to another reference sample. Relative quantification can be achieved through label-based or label-free approaches. Label-based strategies are based on the introduction of stable isotopic labels into glycoproteins or glycopeptides. The glycopeptides from different samples are labelled with different mass tags. These labels change the mass of a glycoprotein or glycopeptide without affecting its analytical properties. Thereby, through MS analysis, the changes in glycoprotein abundance can be obtained by comparing the peak intensity of the differently isotopic labelled glycopeptides. A significant advantage of labelling strategies is that it can simultaneously analyze different samples in a single experiment.

It can also reduce signal variability associated with run-to-run inconsistencies. The isotopic label can be introduced into the samples at different stages of sample preparation. According to different labelling strategies, it can be further divided into chemical labelling, metabolic labelling, and enzymatic labelling methods. In comparison to label-based quantification, label-free quantification does not introduce isotopic labels. The comparisons between samples are based on either mass spectrometric signal intensities of glycopeptides or the number of MS/MS spectra matched to glycopeptides and glycoproteins (spectral counting). Without the need to incorporate a labelling step, the workflow of label-free strategies are generally simpler than label-based strategies. In principle, the fold-change of glycopeptides among different samples can reflect the changes in the glycosylation on glycoproteins. Changes in the non-glycopeptide ratios can reflect changes in glycoprotein abundance at the protein level.

2.2.1 Chemical Labelling

Chemical labelling strategies have been widely employed for glycoprotein quantification in recent years because it can be applied to various kinds of samples. Stable isotopes are introduced at the reactive groups on peptides, such as amino groups and carboxyl groups, through different chemical reactions. After labelling, the isotopically labelled glycopeptides are mixed and analyzed by mass spectrometry. The comparison of signal intensity of light and heavy isotopes labelled glycopeptides affords the relative quantification information. Various kinds of chemical labelling approaches have been reported. In this session, we discuss these strategies according to different chemical labelling methods. Zhang et al. described a method for quantification of N-glycopeptides through succinic anhydride labelling for the first time (Figure 2.2) (Zhang et al. 2003). In this work, they first utilized hydrazide chemistry to selectively enrich glycoproteins. And then, the captured glycoproteins are proteolyzed on the solid support via trypsin and then the glycopeptides were released by peptide N-glycosidase F (PNGase F). Subsequently, the α-amino groups of lysine on glycopeptides were labelled with isotopically light (d0, contains no deuteriums) or heavy (d4, contains four deuteriums) forms of succinic anhydride, which introduced a 4 Da mass difference between the two samples. To illustrate the accuracy of this quantification method, they designed a proof-of-principle experiment. They divided the glycopeptides into two groups with an equal amount and then labelled them with this method. During the

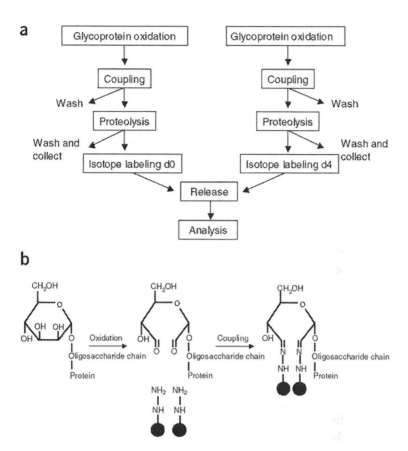

FIGURE 2.2 Schematic diagram of quantitative analysis of N-linked glycopeptides. (A) Strategy for quantitative analysis of glycopeptides. (B) Oxidation of a carbohydrate to an aldehyde followed by covalent coupling to hydrazide resin (Zhang et al. 2003).

MS analysis, the observed ratios between the two samples were generally close to the expected ratios of 1. This result indicated that the accuracy of this quantification approach. Finally, they analyzed the plasma membrane proteins and glycoproteins from human serum, further offering a comprehensive method for quantitative profiling of the glycoproteins in different biological samples.

Succinic anhydride is not universal and gradually replaced by other strategies. For example, dimethyl labelling can be introduced to N-terminal and lysine amino groups of peptides through reductive amination using formaldehyde and cyanoborohydride. By using combinations of isotopic formaldehyde and cyanoborohydride, peptides from different samples can be labelled with a mass difference of a minimum of 4 Da (Boersema et al. 2009). Compared with succinic anhydride, dimethyl labelling can be directly labelled at both lysine amino groups and N-terminal. Moreover, the reaction is fast and highly efficient. Pan et al. described an approach of using a dimethyl labelling strategy to label the glycopeptides for quantification (Pan et al. 2014). In this work, formaldehyde-H_2 and formaldehyde-D_2 were utilized to label peptides at N-terminal and lysine amino groups to introduce 4 Da mass differences between different samples. After PNGase F treatment, the quantification of a glycopeptide can be achieved using the intensity ratio between the heavy and light isotopic forms of the glycopeptides. They analyzed the N-glycoproteome of the human pancreas using this method, and they found several glycoproteins associated with pancreatic cancer showed aberrant N-glycosylation level, including mucin-5AC (MUC5AC), carcinoembryonic antigen-related cell adhesion molecule 5 (CEACAM5), insulin-like growth factor binding protein (IGFBP3), and galectin-3-binding protein (LGALS3BP). In another study, based on the fact that dihydroxy of sialic acid of glycan chains in glycoproteins can be specifically oxidized to the aldehyde in mild periodate concentration while all types of glycan chains can be oxidized in high periodate concentration, Zhang et al. developed a method combining dimethyl labelling with differential oxidation of glycoprotein for determination of sialylation glycan occupancy ratio (Figure 2.3) (Zhang et al. 2014). They utilized

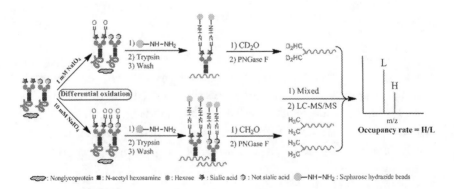

FIGURE 2.3 Schematic diagram of the workflow for the determination of the site-specific N-sialoglycan occupancy rates (Zhang et al. 2014).

1 mM $NaIO_4$ to oxidize the sialic acid exclusively and 10 mM NaIO4 for other glycans. After $NaIO_4$ oxidation, glycoproteins were enriched using hydrazide chemistry, followed by tryptic digestion, dimethyl labelling, and PNGase F digestion. The released deglycosylated peptides were collected for MS analysis. Finally, 496 and 632 site-specific sialylation glycan occupancy ratios on 334 and 394 proteins from hepatocellular carcinoma (HCC) and normal human liver tissues were obtained through this approach. By comparing the sialylation glycan occupancy ratios between HCC and normal human liver tissue, 76 N-sialoglycosites with more than a 2-fold change were determined. Chen et al. used a reverse glycoblotting strategy to specially enrich sialylated glycopeptides and dimethyl labelling for quantification (Chen et al. 2012). Finally, 69 aberrant N-glycosylation sites were identified in sera samples from HCC patients. Compared with sera from liver cirrhosis patients, six glycosylation sites showed elevated levels of aberrancy in HCC patients.

Except for dimethylation, various labelling strategies using different stable isotope reagents have also been proposed, such as stable isotopic acetylation reagent. Qiu et al. used *Sambucus nigra* agglutinin (SNA) to enrich sialylated glycoproteins and labelled glycopeptides with acetylated stable isotope-coding reagents for glycoprotein quantification (Qiu and Regnier 2005). Concretely, the peptides from the control and experimental samples were acetylated at the N-terminus with N-acetoxysuccinamide and deuterated N-acetoxysuccinamide. After being mixed, the mass difference was generated between two different samples. The amount difference in sialylated glycoproteins between the control and experimental samples were obtained by comparing their isotopic labelled glycopeptides. Also, they used concanavalin A (Con A) coupled with SNA to capture sialylated glycopeptides to detect the degree of sialylation within glycoproteins. They used a set of serial lectin columns consisting of a Con A column and an SNA column to select sialylated glycopeptides with complex bi-antennary N-linked, hybrid, and high-mannose glycans. Then these glycopeptides were labelled with the light acetylation reagents. At the same time, they used Con A lectin column alone to enrich glycopeptides containing complex bi-antennary N-linked, hybrid, and high-mannose glycans without regard to sialylation, and these glycopeptides were labelled with heavy acetylation reagents. These glycopeptides selected were mixed, deglycosylated by PNGase F, and fractionated by reversed-phase liquid chromatography (RPLC). The RPLC fractions were then analyzed by electrospray ionization mass spectrometry (ESI-MS). By using this strategy, they quantified the relative amount of sialylation of glycopeptides from human serum enriched by Con A and found that sialylation on various sites in the same protein is independently regulated.

Proteins contain a lot of amino acids that can be labelled. 2-nitrobenzenesulfenyl chloride (NBS) labelling involves the introduction of light and heavy NBS moieties to tryptophan residues, resulting in a 6 Da mass difference between light and heavy NBS-labelled tryptophan peptide (Ou et al. 2006). Ueda et al. (2007) used lectin column chromatography to enrich glycoproteins and labelled glycopeptides using isotopic NBS labelling to quantify glycopeptides. *Lens culinaris* lectin affinity enrichment can separate complex oligosaccharides containing alpha-1,6-linked fucose. They applied this approach to analyze five serum samples derived from patients with lung adenocarcinoma. Thirty-four serum glycoproteins that showed a significant difference in a-1,6-fucosylation level between lung cancer and healthy control were identified as candidate biomarkers. However, NBS-labelled peptides often exhibit low signal intensity compared to other peptides during ionization, resulting in difficulties for relative quantification. Moreover, compared with lysine residues, the efficiency of labelling reaction on tryptophan is lower. Also, the tryptophan residue is one of the least abundant amino acid residues, also making the labelling less efficient. Hence, the NBS-based labelling strategy is not universally applied.

In previous paragraphs, we mainly discussed non-isobaric chemical labelling. Through these methods, glycopeptides in different samples could be quantified by comparing the peak intensities or peak areas of light and heavy labelled peptides at the MS1 level. To analyze multiple samples simultaneously, glycopeptides from different samples are usually labelled with isobaric mass tags, such as isobaric tags for relative and absolute quantification (iTRAQ) and tandem mass tag (TMT) labelling (Kang et al. 2020; Ross et al. 2004). The principle of the isobaric labelling approach is that peptides from different samples can be isobarically labelled. The quantification can be realized by utilizing sequence-specific MS/MS ions. The significant merit of this type of method is that isobaric labelling enables relative quantification for many samples simultaneously in one experiment. Moreover, the labelled peptides were co-eluted as a single composite peak with the same m/z value in the MS1 scan, and the complexity of MS1 is not increased (Li et al. 2020). Kang et al. identified novel signalling pathways involved in the initial release of insulin from pancreatic beta-cells (PBCs) after glucose stimulation using iTRAQ-labelling to assess phosphorylated proteins and sialylated glyco-proteins (Kang et al. 2018). Islets of Langerhans derived from newborn rats with a subsequent 9–10 days of maturation in vitro were stimulated with 20 mM glucose for 0 minute (control), 5 minutes, 10 minutes, and 15 minutes, and then the islets of Langerhans were subjected to quanti-tative phosphoproteomics and sialiomics analysis using iTRAQ-labelling combined with enrichment of phosphorylated peptides and sialylated glycopeptides. They identified six novel activated signalling pathways at 15 minutes glucose-stimulated insulin secretion (GSIS), including agrin interactions and prolactin signalling pathways, which increased the understanding of the molecular mechanism underlying GSIS (Figure 2.4).

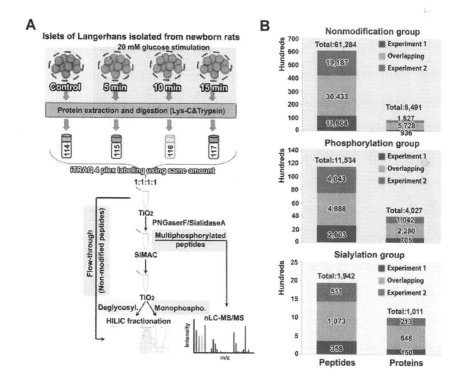

FIGURE 2.4 Experimental set-up and identification overview. (A) Rat pancre-atic islets of Langerhans were stimulated with high glucose (20 mM) for 5, 10, or 15 minutes, and nontreated islets served as control. (B) Summary of the numbers of identified proteins and peptides in the nonmodified, phosphorylated, and N-linked sialylated protein/peptide group, respectively (Kang et al. 2018).

Xue et al. applied iTRAQ-labelling to compare N-glycoproteome changes in 94 serum samples with and without Cryo-thermal therapy treatment (Xue et al. 2016). After tryptic digestion, hydrazide chemistry enrichment, and PNGase F digestion, glycopeptides were labelled by iTRAQ 8-plex kit, and the labelled peptides were analyzed by LC-MS/MS. They quantified 231 highly confident N-glycosylated proteins, and 53 showed significantly discriminated regulatory patterns over the course of time, in which the acute phase response emerged as the most enhanced pathway. iTRAQ and TMT are reliable for glycopeptide quantification because of commercial availability and quality control. The significant advantage of the isobaric labelling is that it enables relative quantification for many samples simultaneously in one experiment, which can improve the analysis throughput but not increase the complexity of the MS1 signal. However, these methods still have some limitations. Compressed ratios have been reported due to impurities at the reporter ion level. And the cost of these commercial reagents is expensive that makes them unsuitable for use during method development or exploratory studies.

2.2.2 Metabolic Labelling

The stable isotope labelling by amino acids in cell culture (SILAC) is a universal metabolic labelling method in quantitative proteomics. Cell lines are grown in media lacking a standard essential amino acid but supplemented with a non-radioactive, isotopically labelled form of that amino acid. After multiple generations of culturing, proteins from different cell lines can be distinguished by isotopically labelled amino acids during MS analysis. Because the stable isotopic tags are incorporated at the earliest stages of sample preparation, it can avoid the quantitative error introduced in the operation process.

SILAC is also a practical technique for quantification of glycoproteome from cells. Clark et al. used the SILAC quantification strategy paired with hydrazide chemistry N-linked glycopeptide enrichment, to examine differentially expressed glycoproteins in cell lines derived from three different states of lung tumorigenesis (Clark et al. 2016). They used light SILAC medium Arg^0/Lys^0 to label HBEC4 cells, and heavy SILAC medium Arg^6/Lys^4 and Arg^{10}/Lys^8 were utilized for A549 cells and HCC827 cells, respectively. After ten cell doublings, the incorporation of the stable isotopes was >98%. The proteins from the three cell lines were harvested, mixed, and digested into peptides, and analyzed by LC-MS/MS. In total, they quantified 118 glycopeptides corresponding to 82 glycoproteins in these three cell lines. Sun et al. investigated a method for the determination of the absolute glycosylation stoichiometry of glycoproteins in the OVCAR-3 cell line (Sun and Zhang 2015). In order to study the glycosite occupancies, they used tunicmycin (TM) to inhibit the overall N-glycosylation occupancies of the cells. Untreated cells were grown in heavy SILAC medium (Arg^{10}/Lys^6), and the treated cells were grown in light SILAC medium (Arg^0/Lys^6). Proteins from these two groups were mixed equally. After tryptic digestion, hydrazide chemistry enrichment, and PNGase F digestion, they analyzed both the deglycosylated peptides and non-glycosylated peptides by LC-MS/MS. Based on the total percentage of glycosylated and non-glycosylated forms of a glycosylation site is 100%, they determined the occupancies at 117 absolute N-glycosylation occupancies in OVCAR-3 cells. Poljak et al. developed an approach to examining glycosylation site occupancy in *Saccharomyces cerevisiae* combined with the SILAC approach with PRM (parallel reaction monitoring)-MS (Poljak et al. 2018). In brief, during the discovery phase, yeast cells were grown in a synthetic drop-out medium, and the harvest proteins were digested with LysC and trypsin proteinases to generate the peptides mixture. The glycopeptides were enriched by ZIC-HILIC, and then glycans were released using Endo H endoglycosidase while retaining a single GlcNAc residue on the glycopeptides. They identified 166 glycosylation sites corresponding to 105 glycoproteins. During the target PRM-MS analysis, the yeast cells were grown in a medium with heavy arginine and lysine isotopes and light arginine and lysine isotopes, respectively. Proteins from these cells were mixed equally. Membrane derived peptides were prepared as described above; glycans were also released with Endo H. In target PRM analysis, among 166 glycopeptides corresponding to 105 glycoproteins measured above, they quantified 62 glycopeptides belonging to 43 glycoproteins in the unenriched samples. Next, they investigated the roles of the Ost3 and Ost6 subunits in the OST function in vivo using this method. Equal amounts of wild-type reference cells grown in the heavy medium, and cells with a deletion in one OST subunit grown in light medium were pooled. The result revealed that 84% of the glycosylation sites were hyperglycosylated in the strain lacking Ost3 protein. In contrast, less severe hyperglycosylation was observed in the strain, with OST6 being overexpressed, with 39% of the glycosylation sites being hyperglycosylated.

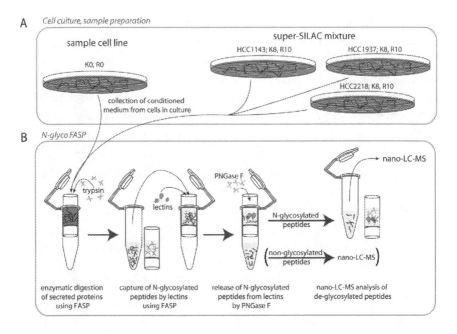

A *Cell culture, sample preparation*

sample cell line

super-SILAC mixture

HCC1143; K8, R10   HCC1937; K8, R10

K0, R0

HCC2218; K8, R10

collection of conditioned
medium from cells in culture

B *N-glyco FASP*

nano-LC-MS

trypsin

PNGase F

lectins

N-glycosylated
peptides

non-glycosylated
peptides   nano-LC-MS

enzymatic digestion
of secreted proteins
using FASP

capture of N-glycosylated
peptides by lectins
using FASP

release of N-glycosylated
peptides from lectins
by PNGase F

nano-LC-MS analysis of
de-glycosylated peptides

FIGURE 2.5   Overview of the experimental workflow. (A) The secretome was collected as a conditioned medium from cell lines in culture. HCC1143, HCC1937, and HCC2218 were also cultured under SILAC conditions to generate a super-SILAC mix as an internal standard. (B) N-glycosylated peptides were enriched using N-glyco FASP (filter aided sample preparation) (Boersema et al. 2013).

However, these methods are limited in their application; most methods can only analyze samples that can be grown in cell culture. To overcome this limitation, Boersema et al. developed a Super-SILAC approach to quantify the N-glycosylated secretome during breast cancer progression and human blood samples (Figure 2.5) (Boersema et al. 2013). First, three breast cancer cell lines were grown in SILAC medium (Arg^{10}/Lys^8), and the mixture was used as an internal standard for quantification. The proteins extracted from different breast cancer tissues could be quantified by comparing their signal intensity with these internal standards. They identified and quantified 1,398 unique N-glycosylation sites in 11 cell lines that are representative of different stages of breast cancer. Five different profiles of glycoprotein dynamics during cancer development were detected. And then, they applied the super-SILAC method to analyze the glycoproteome in plasma. A large number of the previously identified glycopeptides were identified and quantified in body fluid, and several proteins were contained with known roles in breast cancer.

SILAC methods introduced light/heavy isotopic peaks to MS1 scan, but such labelling increases the complexity of MS1 signals, not beneficial for the deep coverage protein identification and quantification. And due to the limited kinds of amino acids with different stable isotopes, it is difficult to simultaneously analyze multiple samples using this method. Also, the cost of reagents is pretty high.

2.2.3 Enzymatic ^{18}O-labelling

^{18}O-labelling, which produces the isotopic label at the C-terminal carboxylic acids of tryptic peptides, is a relatively convenient stable isotope-coding approach. Trypsin catalyzed ^{18}O-labelling leads to the exchange of two ^{16}O atoms for two ^{18}O atoms at the C-terminal carboxyl group of the tryptic peptide (Chahrour et al. 2015). Compared with the above-mentioned chemical labelling approaches, the major advantage of enzymatic ^{18}O-labelling methods is that the labelling step is quite simple because it can be completed during the digestion of glycopeptides. Specifically, for N-glycopeptides analysis, the deglycosylation by PNGase F in $H_2{}^{18}O$ can introduce another ^{18}O at the glycosylation site because the Asn residues are converted to Asp, and the β-carboxyl group of Asp is labelled with one ^{18}O atom at that time (Gonzalez et al. 1992). Kristy et al. were the first to develop a sequential labelling strategy to characterize N-linked glycopeptides (Reynolds et al. 2002). Two samples were parallel hydrolyzed sequentially with Glu-C and PNGase F in $H_2{}^{16}O$ or $H_2{}^{18}O$, respectively. The two steps enzymatic digestion generated a 6 Da difference between a glycopeptide pair. The changes in the glycosylation site occupancy were obtained from the changes in signal intensity ratios of ^{18}O/^{16}O-labelled glycopeptides. They evaluated the method using chicken riboflavin binding protein and showed the good quantitative performance of this method. However, the enzymatic efficiency of Glu-C is not good. To further improve the efficiency of the enzymatic labelled method for quantifying glycopeptides, our group developed a tandem ^{18}O stable isotope labelling (TOSIL) method to quantify the changes of protein expression and individual N-glycosylation site occupancy (Figure 2.6). By employing trypsin to digest the proteins and PNGase F to detach the glycans, the mass of glycopeptides with a single N-glycan site shift +1 and +7 Da after treatment in $H_2{}^{16}O$ and $H_2{}^{18}O$, respectively, which can offer a 6 Da difference with a single glycosylation site (Liu et al. 2010).

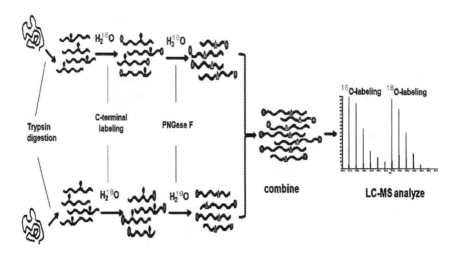

FIGURE 2.6   Analytical strategy of tandem ^{18}O stable isotope labelling (TOSIL) in quantitative proteomics (Liu et al. 2010).

Meanwhile, the non-glycopeptide pairs are separated by 4 Da. Therefore, glycopeptides and non-glycopeptides can be easily distinguished and quantified. This method yielded good linearity in quantitative response within a 10-fold dynamic range with the correlation coefficient $R^2 > 0.99$ and the standard deviation (SD) ranged from 0.06 to 0.21. Using this method, we analyzed N-glycosylation levels in serum from healthy individuals and patients with ovarian cancer. A total of 86 N-glycosylation sites were quantified, and N-glycosylation levels of 56 glycopeptides showed significant change. Shakey et al. developed a similar strategy allowing for relative quantification of glycoproteins in complex biological mixtures (Shakey et al. 2010). The strategy includes periodate oxidation of tryptic digests, solid-phase enrichment of glycopeptides via hydrazide-coupled magnetic beads, and ^{18}O stable isotope labelling catalyzed by trypsin and PNGase F. Subsequently, the glycopeptides were identified and quantified by LC-MS/MS analysis. They identified and quantitated 224 N-glycopeptides representing 130 unique glycoproteins from 20 μL of the undepleted mouse serum samples. This method was able to identify many low-abundance proteins, such as epidermal growth factor receptor (EGFR), insulin-like growth factor-binding protein (IGFBP) complex acid labile chain, and leukemia inhibitory factor receptor.

Enzymatic ^{18}O-labelling also can be combined with other labelling methods. For example, chemical labelling methods can introduce mass tags on peptides for glycopeptide quantification. Meanwhile, the glycosylation site can be labelled by ^{18}O-labelling. Shetty et al. described a lectin-directed tandem labelling quantitative proteomics strategy, combining enzymatic ^{18}O-labelling and acetic anhydride labelling (Shetty et al. 2010). They enriched sialylated glycopeptides by SNA. And then, the glycopeptides were labelled with acetyl (^{1}H$_3$/^{2}D$_3$) groups at the N-terminus by acetic anhydride (^{1}H$_6$/^{2}D$_6$) reagents. After the differentially labelled peptides were enzymatically deglycosylated in heavy water (H$_2$^{18}O), all glycopeptide sequences were identified with high confidence based on the presence of NXT/S motif and a 3 Da mass shift in b or y ions. Further, they implemented this method to investigate the sialylation changes in prostate cancer serum samples and compared them with healthy controls. They identified 45 sialylated glycopeptides. And quantitative results showed that the degree of sialylation was increased in prostate cancer serum samples. In order to quantify multiple samples simultaneously, Ueda et al. combined ^{18}O-labelling and 8-plex iTRAQ labelling together for the discovery of carbohydrate-targeting serum biomarkers (Ueda et al. 2010). They enriched the glycopeptides through lectin column chromatography and labelled the N-glycosylation sites in H$_2$^{18}O during the elution with N-glycosidase. They applied this method to eight sera from lung cancer patients and controls. Finally, 107 glycopeptides were identified in their serum, including A2GL_Asn151, A2GL_Asn290, CD14_Asn132, CO8A_Asn417, C163A_Asn64, TIMP1_Asn30, and TSP1_Asn1049. The change of the affinity to Con A lectin was detected between the lung cancer samples and the controls.

PNGase F treatments were generally utilized to quantify N-glycosylation site occupancy. However, the spontaneous deamidation of asparagine can occur during sample preparation, which may overestimate the glycosylation occupancy during PNGase F treatments for the glycosylation site (Wright 1991). Therefore, incorporation of ^{18}O at the glycosylation site is also beneficial for identifying the glycosylation site. Compared with chemical labelling methods, enzymatic ^{18}O-labelling methods are more convenient and simpler in operation, but incomplete labelling and back exchange should be considered and carefully controlled in experiments.

2.2.4 Label-Free Relative Quantification

Label-free quantification can be divided into two groups: (1) signal intensity measurement based on precursor ion spectra and (2) spectral counting, which is based on counting the number of peptides assigned to a protein in an MS2 experiment. Compared with labelling methods, label-free methods usually have a simple workflow because labelling methods need extra reagents to modify the peptides. Additionally, the number of samples analyzed in a single experiment is limited, and some labelling strategies cannot be easily applied to all types of samples. Label-free methods have been widely used in glycoproteomics in recent years, especially for large-scale sample analysis (Cox et al. 2014; Esquivel et al. 2016).

To meet the different needs of glycoproteome quantification, several acquisition methods have been employed. Data-dependent acquisition (DDA) mode is the most widely adopted one. During this acquisition mode, it automatically selects precursor ions detectable at the full scan level in a given order. Label-free quantification is performed at the full MS scan level by integrating the area of the LC peak from an extracted ion chromatogram of the precursor mass corresponding to the given peptides. Combined with multi-step separation and multi-enzymatic digestion (Lys-C, trypsin, Glu-C, and PNGase F), Jiang et al. characterized the glycosylation occupancy and the active site in the TNK-Tissue plasminogen activator based on peptide-ion intensity to quantify glycopeptides (Jiang et al. 2010). They found that the differences of TNK-tPA in glycosylation occupancy and chain cleavage at the activation site of the enzyme between the innovator and follow-on products are also different. Hu et al. performed an integrated proteomic and glycoproteomic analysis of 83 prospectively collected high-grade serous ovarian carcinoma (HGSC) and compared with 23 non-tumor tissues using the label-free method (Hu et al. 2020). Through quantitative glycoproteome analysis, they revealed the tumor-specific glycosylation differences and proved the altered glycosylation was correlated with the changes of glycosylation enzymes. Their results indicated that it is possible to distinguish pathological outcomes of ovarian tumors from non-tumors using glycoproteins, and glycoproteins also can be utilized to classify the tumor clusters. The label-free method was especially utilized for low microgram of proteins and low abundance of glycoprotein quantification, because no additional reagents and extra sample preparation steps were introduced. Huang et al. developed an integrated spin tip-based approach for sensitive and quantitative profiling of region-resolved in vivo brain glycoproteome. All the steps of glycoprotein enrichment, digestion, deglycosylation, and desalting were in a single spin tip device (Huang et al. 2019). More than 200 N-glycosites were identified from only 0.5 mm^2 (20 μm thickness) of the brain section (around 1 μg proteins). More than 1,000 N-glycosites were identified when the area was increased to 15 mm^2 (around 25 μg proteins). They found 26 N-glycosites were overexpressed in the isocortex, hippocampus, thalamus, and hypothalamus. 26, 12, 7, and 5 glycosites were uniquely quantified in each region. The characteristic of label-free methods is that it can generate large volumes of data, which need rigorous statistical assessment for accurate data processing and interpretation. Therefore, effective algorithm models and software tools were developed. Mayampurath et al. presented a novel analysis of variance (ANOVA)-based mixed-effects model for a label-free quantification method to quantify the site-specific glycosylation of glycoproteins. The abundance of a glycopeptide is broken down into several fixed and random effects. The fixed terms account for variation at the global glycoprotein level, whereas the random effects indicate changes at either the site or glycan level. Through hypothesis testing, they showed site-specific variations (i.e., differential increase or decrease in expression at a particular site and glycan or across all sites with glycans attached) between healthy and disease data sets could be obtained. They used this method for the discovery of esophageal cancer glycosylated biomarkers (Mayampurath et al. 2014). Vitronectin was consistently observed to show aberrant site-specific glycosylation.

Recently, with the development of quantitative proteomics, the data-independent acquisition (DIA) mode has received more and more attention because it can completely record the fragment ion spectra of all the analytes in samples for which the precursor ions are within a predetermined m/z versus retention time window (Liu et al. 2013). Particularly, during glycoproteome quantitative analysis, many glycopeptides are overlooked and not selected for MS/MS fragmentation in DDA experiments because of the low concentration of glycopeptides within proteolytic mixtures and poor ionization efficiency. Therefore, DIA-MS strategies also have been employed for the identification and quantification of glycoproteome (Xu et al. 2015; Yeo et al. 2016). Xu et al. developed a method with SWATH-MS to enable automated measurement of site-specific occupancy at glycosylation sites (Xu et al. 2015). After endoglycosidase H digest leaving glycosylated Asn with a single GlcNAc, protease (trypsin or Asp-N) digestion, and MS analysis, the N-glycosylation site occupancy could be determined by the abundance of the GlcNAc-modified peptide as a fraction of the sum of the GlcNAc-modified and unmodified versions of the same peptide. They used ratiometric analysis of deglycosylated peptides and the total intensities of all peptides from the corresponding proteins to relative quantification of site-specific glycosylation occupancy between yeast strains with various isoforms of oligosaccharyltransferase. They also analyzed the glycosylation occupancy in human salivary glycoproteins and identified the deglycosylated form of 16 glycosylation sites from 8 glycoproteins. DIA-MS can improve the reproducibility of quantification methods for its ability to select low-lying precursor masses. However, there are still several challenges. DIA-MS analysis relies on a spectral library of sufficient proteome coverage. To achieve a comprehensive library with deep-coverage, DDA-MS analysis of the same sample at the same time of instrument is required, which is time-consuming and sample-consuming (Tsou et al. 2015). Moreover, because the connection of precursor ions and fragments was lost, the quantification of post-translationally modified peptides is also a challenge.

Label-free methods usually have a simple workflow, and it can be utilized for multiple sample quantification in a single experiment. Also, it is suitable for all types of samples. However, reproducibility and data processing should be considered carefully in label-free methods because each sample is handled separately, starting from sample preparation to the final detection.

2.3 ABSOLUTE QUANTIFICATION

The approaches mentioned above all dealt with the relative quantification of proteins across samples. The absolute quantities or concentrations of glycoproteins cannot be obtained by these methods. Absolute quantification information is important for the discovery of diagnostic markers, predictive markers, and prognostic markers. Targeted mass spectrometry techniques, specifically multiple reaction monitoring (MRM) on triple quadruple instruments and PRM on orbitraps, have been adopted in glycoproteomics due to more precise quantification, high analytical reproducibility, better signal-to-noise ratios, and increased dynamic range (Vidova and Spacil 2017). In these analysis approaches, a sequential selection of peptide specific precursor ions in the first mass analyzer and a characteristic fragment ion in the second mass analyzer upon fragmentation in the collision cell were combined, with stable isotope-labelled peptide as a reference to quantify the absolute amount of targeted glycoproteins (Stahl-Zeng et al. 2007).

For absolute quantification, the proteins with known concentrations are used as an internal standard. The stable isotope labelled peptides are a desirable choice because they could be co-eluted with the target peptides at the same time, which can provide reliable absolute quantitative results. Meanwhile, the technology for the synthesis of stable isotope labelled peptides is mature. But for glycoproteome quantification, isotope labelled deglycosylation peptides were widely utilized as the internal standard due to the synthesis of isotope labelled intact glycopeptide standards remains difficult. And then, the prepared stable isotope-labelled internal standards of known absolute quantity were added to a sample. The absolute amounts of peptides were determined by comparing their MS signal intensities with those standards. Shiao et al. developed a method that coupled the on-bead enzymatic protein digestion method to determine IgG subclass and glycosylation (Figure 2.7) (Shiao et al. 2020). They first utilized protein G to capture IgG, and the captured proteins were digested on beads. A stable isotope-labelled IgG was incorporated into the serum

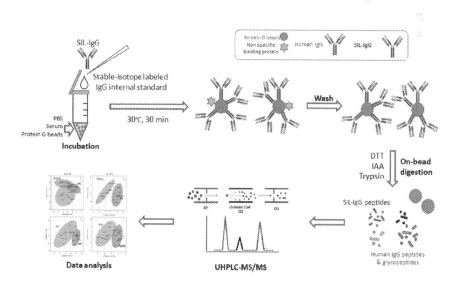

FIGURE 2.7 Workflow of the ultra-high-performance liquid chromatography (UHPLC)-MS/MS method for analyzing human serum IgG profiles.

as an internal standard and incubated together to achieve accurate quantification. This method was applied to quantify IgG from patients with autoimmune pancreatitis (AIP), and pancreatic ductal adenocarcinoma (PDAC), and samples from healthy people as controls. They measured that IgG concentrations are around 10 μg/μL in most of the individuals. The concentrations of IgG1, IgG2, IgG3, and IgG4 are 3.19–10.2 μg/μL, 1.23–6.63 μg/μL, 0.16–1.94 μg/μL, and 0.03–1.33 μg/μL, respectively (R^2>0.995). They found that distinct IgG patterns were discovered among the groups, and seven glycopeptides showed high potential in differentiating AIP and PDAC. MRM-MS analyses are usually combined with enrichment strategies to improve the specificity and accuracy of quantification for target glycopeptides. Zhao et al. described a method that used lectin enrichment (core fucosylation glycoproteins captured by Lens Culinaris Lectin (LCH) sepharose 4B) combined with the MRM-MS technique to quantify core fucosylated glycoproteins (Zhao et al. 2011). They used Endo F3-catalyze to release glycans. MRM-MS analysis was introduced to obtain site-specific quantification information of core fucosylated peptides. To illustrate the feasibility of the quantification method, the core fucosylated peptide of target proteins in clinical serum was quantified and compared as a preliminary demonstration. Due to the extraordinary complexity of the serum proteome, it is a challenge to direct quantification of glycoproteins in serum. Cao et al. described a non-glycopeptide-based mass spectrometry (NGP-MS) strategy (Cao et al. 2019). In brief, glycoproteins were first captured on solid beads through hydrazide chemistry. After thorough washing, the non-glycopeptides of the captured glycoproteins were released through a trypsin digestion and analyzed by LC-MS/MS. Finally, the selected NGPs were synthesized for the MRM assay. They applied this method for HCC glycoprotein biomarker discovery from 20 HCC serum mixtures and 20 normal serum mixtures. A total of 97 glycoprotein candidates were preliminarily screened and submitted for absolute quantification with NGP-based stable-isotope-labelled (SID)-MRM in the individual samples of 38 HCC serum and 24 normal controls. Finally, 21 glycoproteins were absolutely quantified. They found three glycoproteins, beta-2-glycoprotein 1 (APOH), alpha-1-acid glycoprotein 2 (ORM2), and complement C3 (C3) could be utilized to distinguish the HCC patients and healthy people.

When faced with complex sample detection, the specificity and sensitivity of MRM technology still cannot meet the requirements. So PRM

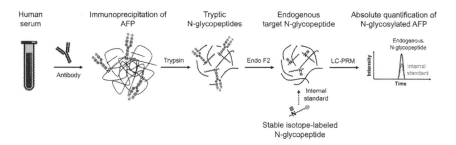

FIGURE 2.8 Analytical workflow for the AQUA of N-glycosylated AFP with stable isotope-labelled *N*-glycopeptide by LC-PRM.

methods have been developed. The third quadrupole of a triple quadrupole is substituted with a high-resolution and accurate mass analyzer to permit the parallel detection of all target product ions in one, concerted high-resolution mass analysis. Kim et al. developed an endoglycosidase-assisted absolute quantification (AQUA) method to measure N-glycosylated AFP levels in serum using LC-MS/MS-PRM (Figure 2.8) (Kim et al. 2020). They used a ^{13}C- and ^{15}N-labelled N-glycopeptides of a simple glycosylated structure (peptide + HexNAc) to directly determine the N-glycosylation levels of AFP in human serum. The isotope-labelled N-glycopeptide of AFP was spiked as an internal standard into serum after trypsin and Endo F2 treatment. The N-glycopeptides of the target and internal standard were analyzed by LC-PRM MS for AQUA. This is also the first report that an isotope-labelled glycopeptide was used as an internal standard. The N-glycosylation levels of AFP were measured in HCC patients and in healthy donors, both cases using this method. They observed N-glycosylated AFP levels (3.7–6.8 ng/mL), which are more than two times higher than the LLOQ (1.95 ng/mL) with less than 10% of CV (coefficient of variation) value by LC-PRM. They also found that N-glycosylation levels of AFP showed better correlation with tumor grades of HCC patients compared to total AFP levels from the ELISA assay (enzyme-linked immunosorbent assay).

Although MRM analyses are highly specific, monitoring specific targets and transitions over the entire chromatographic timeframe severely reduces the number of analytes that may be quantified. In addition, MRM analysis requires the synthesis of stable isotope-labelled internal standards, and the cost of these peptides is high.

2.4 CONCLUSION

The key to the eventual utility and understanding of glycosylation in biology and disease diagnosis lies in the ability to quantitate protein glycosylation. Recent researches have achieved high-throughput and highly sensitive methods that can accurately quantify glycosylation site occupancy and analyze the relationship between the changes in glycosylation site occupancy and protein expression, which performed either at the relative or the absolute level. However, there are also some notable questions. First, for labelling methods, the efficiency of labelling reagents should be considered carefully. Second, the reproducibility of label-free methods relies on the state of instruments. More robust and reproducible LC-MS/MS methods should be developed to increase quantification accuracy. Third, it is critical to developing supporting software. Fourth, due to complexity of glycoproteins and glycopeptides, more effective and selective methods must be developed. With the sufficient development of analytical methods in quantitative glycoproteomics research, the host of biological problems in glycobiology could be solved in the future (Sun et al. 2016).

REFERENCES

Boersema, P. J., R. Raijmakers, S. Lemeer, et al. 2009. Multiplex peptide stable isotope dimethyl labeling for quantitative proteomics. *Nature Protocols* 4(4): 484–94.

Boersema, P. J., T. Geiger, J. R. Wisniewski, et al. 2013. Quantification of the N-glycosylated secretome by super-SILAC during breast cancer progression and in human blood samples. *Molecular & Cellular Proteomics* 12(1): 158–71.

Cao, W. Q., B. Y. Jiang, J. M. Huang, et al. 2019. Straightforward and highly efficient strategy for hepatocellular carcinoma glycoprotein biomarker discovery using a nonglycopeptide-based mass spectrometry pipeline. *Analytical Chemistry* 91(19): 12435–43.

Chahrour, O., D. Cobice, J. Malone. 2015. Stable isotope labelling methods in mass spectrometry-based quantitative-proteomics. *Journal of Pharmaceutical and Biomedical Analysis* 113: 2–20.

Chen, R., F. J. Wang, Y. X. Tan, et al. 2012. Development of a combined chemical and enzymatic approach for the mass spectrometric identification and quantification of aberrant N-glycosylation. *Journal of Proteomics* 75(5): 1666–74.

Clark, D. J., Y. P. Mei, S. S. Sun, et al. 2016. Glycoproteomic approach identifies KRAS as a positive regulator of CREG1 in non-small cell lung cancer cells. *Theranostics* 6(1): 65–77.

Cox, J., M. Y. Hein, C. A. Luber, et al. 2014. Accurate proteome-wide label-free quantification by delayed normalization and maximal peptide ratio extraction, termed MaxLFQ. *Molecular & Cellular Proteomics* 13(9): 2513–26.

Engle, D. D., H. Tiriac, K. D. Rivera, et al. 2019. The glycan CA19-9 promotes pancreatitis and pancreatic cancer in mice. *Science* 364(6446): 1156–62.

Esquivel, R. N., S. Schulze, R. Xu, et al. 2016. Identification of *Haloferax volcanii* pilin N-glycans with diverse roles in pilus biosynthesis, adhesion, and microcolony formation. *Journal of Biological Chemistry* 291(20): 10602–14.

Gonzalez, J., T. Takao, H. Hori, et al. 1992. A method for determination of N-glycosylation sites in glycoproteins by collision-induced dissociation analysis in fast-atom-bombardment mass-spectrometry-identification of the positions of carbohydrate-linked asparagine in recombinant alpha-amylase by treatment with peptide N-glycosidase F in O-18-labeled water. *Analytical Biochemistry* 205(1): 151–8.

Hu, Y. W., J. B. Pan, P. Shah, et al. 2020. Integrated proteomic and glycoproteomic characterization of human high-grade serous ovarian carcinoma. *Cell Reports* 33(3): 108276.

Huang, P. W., H. J. Li, W. N. Gao, et al. 2019. A fully integrated spin tip-based approach for sensitive and quantitative profiling of region-resolved in vivo brain glycoproteome. *Analytical Chemistry* 91(14): 9181–9.

Jiang, H. T., S. L. Wu, B. L. Karger, et al. 2010. Characterization of the glycosylation occupancy and the active site in the follow-on protein therapeutic: TNK-tissue plasminogen activator. *Analytical Chemistry* 82(14): 6154–62.

Kang, T., P. Jensen, H. G. Huang, et al. 2018. Characterization of the molecular mechanisms underlying glucose stimulated insulin secretion from isolated pancreatic beta-cells using post-translational modification specific proteomics (PTMomics). *Molecular & Cellular Proteomics* 17(1): 109–24.

Kang, T., B. B. Boland, P. Jensen, et al. 2020. Characterization of signaling pathways associated with pancreatic beta-cell adaptive flexibility in compensation of obesity-linked diabetes in db/db mice. *Molecular & Cellular Proteomics* 19(6): 971–93.

Kim, K. H., S. Y. Lee, D. G. Kim, et al. 2020. Absolute quantification of N-glycosylation of alpha-fetoprotein using parallel reaction monitoring with stable isotope-labeled N-glycopeptide as an internal standard. *Analytical Chemistry* 92(18): 12588–95.

Li, J. M., J. G. Van Vranken, L. P. Vaites, et al. 2020. TMTpro reagents: a set of isobaric labeling mass tags enables simultaneous proteome-wide measurements across 16 samples. *Nature Methods* 17(4): 399–404.

Liu, Z., L. Cao, Y. F. He, et al. 2010. Tandem O-18 stable isotope labeling for quantification of N-glycoproteome. *Journal of Proteome Research* 9(1): 227–36.

Liu, Y. S., R. Huttenhain, S. Surinova, et al. 2013. Quantitative measurements of N-linked glycoproteins in human plasma by SWATH-MS. *Proteomics* 13(8): 1247–56.

Lu, H. J., Y. Zhang, P. Y. Yang. 2016. Advancements in mass spectrometry-based glycoproteomics and glycomics. *National Science Review* 3(3): 345–64.

Mayampurath, A., E. W. Song, A. Mathur, et al. 2014. Label-free glycopeptide quantification for biomarker discovery in human sera. *Journal of Proteome Research* 13(11): 4821–32.

Ou, K., D. Kesuma, K. Ganesan, et al. 2006. Quantitative profiling of drug-associated proteomic alterations by combined 2-nitrobenzenesulfenyl chloride (NBS) isotope labeling and 2DE/MS identification. *Journal of Proteome Research* 5(9): 2194–206.

Pan, S., R. Chen, Y. Tamura, et al. 2014. Quantitative glycoproteomics analysis reveals changes in N-glycosylation level associated with pancreatic ductal adenocarcinoma. *Journal of Proteome Research* 13(3): 1293–306.

Pinho, S. S., C. A. Reis. 2015. Glycosylation in cancer: mechanisms and clinical implications. *Nature Reviews Cancer* 15(9): 540–55.

Poljak, K., N. Selevsek, E. Ngwa, et al. 2018. Quantitative profiling of N-linked glycosylation machinery in yeast saccharomyces cerevisiae. *Molecular & Cellular Proteomics* 17(1): 32–44.

Qiu, R. Q., F. E. Regnier. 2005. Use of multidimensional lectin affinity chromatography in differential glycoproteomics. *Analytical Chemistry* 77(9): 2802–09.

Reynolds, K. J., X. D. Yao, C. Fenselau. 2002. Proteolytic O-18 labeling for comparative proteomics: evaluation of endoprotease Glu-C as the catalytic agent. *Journal of Proteome Research* 1(1): 27–33.

Ross, P. L., Y. L. N. Huang, J. N. Marchese, et al. 2004. Multiplexed protein quantitation in saccharomyces cerevisiae using amine-reactive isobaric tagging reagents. *Molecular & Cellular Proteomics* 3(12): 1154–69.

Ruhaak, L. R., G. G. Xu, Q. Y. Li, et al. 2018. Mass spectrometry approaches to glycomic and glycoproteomic analyses. *Chemical Reviews* 118(17): 7886–930.

Shakey, Q., B. Bates, J. A. Wu. 2010. An approach to quantifying N-linked glycoproteins by enzyme-catalyzed O-18(3)-labeling of solid-phase enriched glycopeptides. *Analytical Chemistry* 82(18): 7722–8.

Shetty, V., Z. Nickens, P. Shah, et al. 2010. Investigation of sialylation aberration in N-linked glycopeptides by lectin and tandem labeling (LTL) quantitative proteomics. *Analytical Chemistry* 82(22): 9201–10.

Shiao, J. Y., Y. T. Chang, M. C. Chang, et al 2020. Development of efficient on-bead protein elution process coupled to ultra-high performance liquid chromatography-tandem mass spectrometry to determine immunoglobulin G subclass and glycosylation for discovery of bio-signatures in pancreatic disease. *Journal of Chromatography A* 1621: 461039.

Stadlmann, J., J. Taubenschmid, D. Wenzel, et al. 2017. Comparative glycoproteomics of stem cells identifies new players in ricin toxicity. *Nature* 549(7673): 538–42.

Stahl-Zeng, J., V. Lange, R. Ossola, et al. 2007. High sensitivity detection of plasma proteins by multiple reaction monitoring of N-glycosites. *Molecular & Cellular Proteomics* 6(10): 1809–17.

Stowell, S. R., T. Z. Ju, R. D. Cummings. 2015. Protein glycosylation in cancer. In *Annual Review of Pathology: Mechanisms of Disease*, edited by A. K. Abbas, S. J. Galli, P. M. Howley,10: 473–510.

Sun, S. S., H. Zhang. 2015. Large-scale measurement of absolute protein glycosylation stoichiometry. *Analytical Chemistry* 87(13): 6479–82.

Sun, S. S., P. Shah, S. T. Eshghi, et al. 2016. Comprehensive analysis of protein glycosylation by solid-phase extraction of N-linked glycans and glycosite-containing peptides. *Nature Biotechnology* 34(1): 84–8.

Tsou, C. C., D. Avtonomov, B. Larsen, et al. 2015. DIA-umpire: comprehensive computational framework for data-independent acquisition proteomics. *Nature Methods* 12(3): 258–64.

Ueda, K., T. Katagiri, T. Shimada, et al. 2007. Comparative profiling of serum glycoproteome by sequential purification of glycoproteins and 2-nitro-benzenesulfenyl (NBS) stable isotope labeling: a new approach for the novel biomarker discovery for cancer. *Journal of Proteome Research* 6(9): 3475–83.

Ueda, K., S. Takami, N. Saichi, et al. 2010. Development of serum glycoproteomic profiling technique; simultaneous identification of glycosylation sites and site-specific quantification of glycan structure changes. *Molecular & Cellular Proteomics* 9(9): 1819–28.

Varki, A. 2017. Biological roles of glycans. *Glycobiology* 27(1): 3–49.

Vidova, V., Z. Spacil. 2017. A review on mass spectrometry-based quantitative proteomics: targeted and data independent acquisition *Analytica Chimica Acta* 964: 7–23.

Wright, H. T. 1991. Nonenzymatic deamidation of asparaginyl and glutaminyl residues in proteins. *Critical Reviews in Biochemistry and Molecular Biology* 26(1): 1–52.

Xu, Y., U. M. Bailey, B. L. Schulz. 2015. Automated measurement of site-specific N-glycosylation occupancy with SWATH-MS. *Proteomics* 15(13): 2177–86.

Xue, T., P. Liu, Y. Zhou, et al. 2016. Interleukin-6 induced "acute" phenotypic microenvironment promotes Th1 anti-tumor immunity in cryo-thermal therapy revealed by shotgun and parallel reaction monitoring proteomics. *Theranostics* 6(6): 773–94.

Yeo, K. Y. B., P. K. Chrysanthopoulos, A. S. Nouwens, et al 2016. High-performance targeted mass spectrometry with precision data-independent acquisition reveals site-specific glycosylation macroheterogeneity. *Analytical Biochemistry* 510: 106–13.

Zhang, H., X. J. Li, D. B. Martin, et al. 2003. Identification and quantification of N-linked glycoproteins using hydrazide chemistry, stable isotope labeling and mass spectrometry. *Nature Biotechnology* 21(6): 660–6.

Zhang, Z., Z. Sun, J. Zhu, et al. 2014. High-throughput determination of the site-specific N-sialoglycan occupancy rates by differential oxidation of glycoproteins followed with quantitative glycoproteomics analysis. *Analytical Chemistry* 86(19): 9830–7.

Zhao, Y., W. Jia, J. F. Wang, et al. 2011. Fragmentation and site-specific quantification of core fucosylated glycoprotein by multiple reaction monitoring-mass spectrometry. *Analytical Chemistry* 83(22): 8802–9.

CHAPTER 3

Qualitative and Quantitative Methods for N-Glycans in N-Glycomics

Shifang Ren and Haojie Lu

CONTENTS

DOI: 10.1201/9781003185833-3

3.1 INTRODUCTION

Like the definition of genome and proteome, N-glycome is generally defined as the general term of all N-glycans in a single organism or a cell type, and N-glycomics is systems-level analysis of the N-glycome in biological systems (Cummings and Pierce 2014). The N-glycome generally refers to all N-glycans released from glycoproteins, glycolipids and other glycoconjugates of organisms (animals, plants, bacteria, etc.), and this chapter mainly focuses on the glycome of mammalian glycoproteins.

N-glycan is involved in many biological behaviours, such as cell development, differentiation, tumour metastasis, immune response (Helenius and Aebi 2001). Glycans have extensive structural heterogeneity. On the basis of genomics and proteomics, the development of glycomics research is conducive to the in-depth and complete understanding of the complex life activities, so as to explain the full function of genes in the organism. As the alteration of N-glycans has been shown associated with many pathological events, systemic approaches based on N-glycomics might significantly contribute to more effective biomarker and therapy development for different kinds of diseases.

The complexity of N-glycan structure is vast as no synthetic template is found in the glycosylation processing and branched structure can be formed with multiple possible glycosidic linkages. These characteristics pose a big analytical challenge for qualitative and quantitative N-glycomics. The rapid development of mass spectrometry (MS) methods for N-glycomic analysis has made it possible to comprehensively obtain glycan structure and glycan-level information in biological samples. The general analytical strategy based on MS for N-glycomics analysis was shown in Figure 3.1, usually including the steps of glycan release, glycan purification, glycan derivatization, matrix-assisted laser desorption/ionization-mass spectrometry (MALDI-MS) or liquid chromatography-electrospray ionization-mass spectrometry (LC-ESI-MS) analysis and data analysis. Over the past decades, considerable progress has been made on the analytical methods for N-glycans from biological sources. This chapter

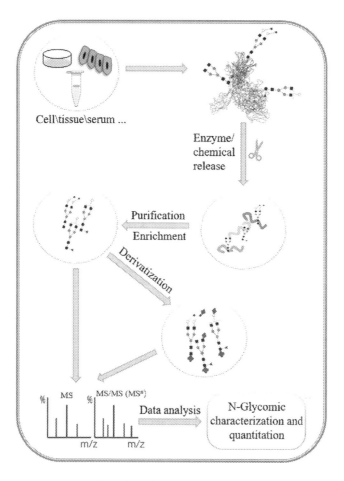

FIGURE 3.1   General MS-based N-glycome analysis strategies.

describes various approaches for qualitative and quantitative analysis of N-glycome, focusing on the methods of sample preparation, identification as well was quantitation. In practice, the specific method is usually chosen depending on the purpose of the study. Whatever N-glycomics analysis, qualitative or quantitative, they mostly involve the pretreatment processes of glycan release, purification, enrichment, separation and derivatization, etc. First of all, we describe some common techniques and methods involve in sample pretreatment. Then the qualitative and quantitative MS analysis methods with the optimization and selective combination of those sample preparation steps are introduced.

3.2 GENERAL SAMPLE PRETREATMENT METHODS PRIOR TO MS ANALYSIS

3.2.1 Release of N-Glycome

The efficient release of N-glycome from glycoproteins or glycopeptides is critical in glycomics analysis. There are several major approaches for releasing N-glycome including enzymatic and chemical methods, as shown in Table 3.1.

3.2.1.1 Enzymatic Release Methods

Peptide-N-glycosidases F and A (PNGase F and PNGase A) (Altmann et al. 1995), endoglycosidases F1-F3 and H (Endo F1-F3 and Endo H) (Maley et al. 1989) have been utilized frequently for N-glycome release. Among them, PNGase F is a universally used one, especially for mammalian glycoproteins. It can specifically cleave the amide bond between the side chain of the Asn residue at the polypeptide chain and the reducing end of N-glycan, producing glycosylamines which would be then hydrolyzed to form reducing N-glycans. However, PNGase F is not effective at releasing N-linked glycan with α (1,3)-linked core fucose that is a typical feature of invertebrate and plant N-glycans (Staudacher et al. 1999). Alternatively, PNGase A, works for all N-glycans regardless of the presence of α (1,3)-linked core fucose residues. The limitation of PNGase A is the lower efficiency in N-glycan release from glycoproteins than that from glycopeptides and risk of interfering with sample analysis because it is a glycoprotein itself. Different from PNGase F and A, Endo F1-F3 and Endo H are frequently used to cleave between the two GlcNAc residues in the core region with one GlcNAc remaining bound to the protein. These endoglycosidases have specificity for different types of glycans. Endo H and Endo F1 can release high-mannose and some hybrid-type oligosaccharides but not complex-type oligosaccharides. Endo F2 can release high-mannose and bi-antennary complex-type oligosaccharides, and Endo F3 cleaves bi-antennary and tri-antennary complex-type oligosaccharides, especially the core-fucosylated structures (Zhang et al. 2018). In addition to the enzyme we mentioned above, some newly discovered enzymes, such as PNGase H$^+$ (Guo et al. 2020) and PNGase Ar (Yan et al. 2018), were also reported for N-glycomics analysis. They have potential as a valuable addition to the currently existing PNGases used for the analysis of N-glycans, although further studies might be necessary to assess their practical applicability.

TABLE 3.1   Some Major Methods for Cleavage of N-Glycan

Type	Method	Cleavage	Pros and Cons
Enzyme Digestion	PNGase F	Cleave the amide bond between the side chain of the Asn residue at the polypeptide chain and the reducing end of N-glycan	Mild conditions, stable and reliable. Expensive; may not effective for release N-glycan with α 1.3 core fucose.
	PNGase A		Works for all N-glycans. Suitable for enzymatic reaction in acidic conditions. Lower efficiency in N-glycan release from glycoproteins than that from glycopeptides. Risk of interfering with sample analysis because it is glycoprotein itself.
	Endo F1	Cleave between the two core GlcNAc residues	Less effective for N-glycans with core fucosylation
	Endo F2		Effective for bi- and tri-antennary complex N-glycan, but less effective for high-mannose N-glycans
	Endo F3		Effective for bi- and tri-antennary complex N-glycans, especially those with core fucosylation
	Endo H		Effective for high-mannose and hybrid N-glycans
	PNGase H+	Cleave the amide bond between the side chain of the Asn residue at the polypeptide chain and the reducing end of N-glycan	Surprisingly low pH optimum of pH 2.5. Suitable for release of N-glycans from native glycoproteins or those with core fucosylation in acidic conditions. Difficulties in optimizing expression and purification of the enzyme at larger scales.

(*Continued*)

TABLE 3.1 (*Continued*)   Some Major Methods for Cleavage of N-Glycan

Type	Method	Cleavage	Pros and Cons
Chemical Release	Hydrazinolysis	Cleaves amide bonds between peptide and N-glycosidic bonds	Cheap. Release in the glycosylamines with N-acetyl moieties of glucosamine lost which must be re-N-acetylated. Issues of safety as well as artefact.
	Alkaline hydrolysis (alkaline or alkaline borohydride)	Cleave between Asn residue and the reducing end of N-glycans.	Cheap. Not widely accepted for N-glycomic studies.
	Oxidation (NaClO)	Release all types of free reducing N-glycans and O-glycan acids from glycoproteins.	Cheap and rapid. Suitable for producing large quantities of natural glycans from unprocessed biological samples. Potential loss of core GlcNAc due to over oxidation; low recovery of released N-glycans from small amounts (<500 mg) of sample.

For enzymatical release method, a number of studies have been used to shorten reaction time and reduce the consumption, including microwave radiation (Zhou et al. 2012), high pressure-cycling technology (PCT) (Szabo et al. 2010) and enzyme immobilization (Ren et al. 2014). It was shown that microwave irradiation can completely remove glycans within 10 minutes from monoclonal antibodies and up to 1 hour from other glycoprotein standards (Sandoval et al. 2007; Zhou et al. 2012). Highly efficient release of N-glycans within 5 minutes was achieved in a recently reported work by employing the oriented immobilization of PNGase F on magnetic particles in combination with microwave-assisted enzymatic digestion techniques (Zhang et al. 2020a).

3.2.1.2 Chemical Release Methods

Some traditional chemical release methods are also used for releasing N-glycans from Asn in the peptide backbone, mainly including hydrazinolysis or alkaline hydrolysis. Hydrazinolysis has been a traditional method for the complete release of N-glycans in the early days (Takasaki et al. 1982). In hydrazinolysis method, anhydrous hydrazine reacts at the link between the glycan and peptide backbone as well as the acetamido groups of monosaccharide residues such as GlcNAc, and thus N-acetyl moieties of glucosamine could be lost during the release. Thus, re-acetylation after release is included in this method which will introduce excess salt and require further desalting step. Meanwhile, the reagent of anhydrous hydrazine is highly toxic and explosive which must be used with caution and in small-scale (e.g., 0.2–1 mL) reaction.

Alkaline hydrolysis is another way to liberate glycans from glycoprotein(s) by employing ammonium hydroxide/carbonate or sodium hydroxide in conjunction with sodium borohydride (reductive β-elimination) (Figl and Altmann 2018) or not (non-reductive β-elimination) (Yuan et al. 2014). Non-reductive β-elimination can release N-glycans in their nonreduced forms under different reaction conditions, favouring subsequent or one-pot labelling of the reducing end using different tags such as fluorescent or ultraviolet (UV) tags. The released nonreduced glycan is prone to degradation from the reducing core monosaccharide under alkaline conditions known as "peeling reaction". To prevent "peeling" reactions, alkaline hydrolysis can be performed in the presence of excess reducing agent such as sodium borohydride and thus the reducing ends of the glycans are reduced into base stable alditols, which is known as reductive β-elimination method. The primary limitation of alkaline β-elimination is the high sodium salt content remaining after reaction which interferes with downstream mass spectrometry analysis.

Another chemical strategy termed as oxidative release of natural glycans (ORNG) which uses household bleach (NaClO) to release all types of glycans from glycoproteins is a latest reported novel chemical release method (Song et al. 2016). This is a cheap and rapid method to produce large quantities of natural glycans from unprocessed biological samples which is especially practical for functional glycomics analysis (Song et al. 2016). It was later reported that absolute recovery of N-linked glycans was significantly lower with ORNG method comparing with typical PNGase F protocol when dealing with small amounts (<500 mg) of sample, while the relative glycan composition of released N-linked glycans are similar (Fischler and Orlando 2019).

3.2.2 Purification, Enrichment and Separation of N-Glycome

After glycan release, glycans are in the presence of interfering components such as enzymes, proteins, peptides and some other chemicals involved in sample preparation that may suppress glycan ionization and detection. To remove excess contaminants and concentrate the glycans to desirable concentrations, proper glycan purification and enrichment of the glycans from the mixtures is essential for the success of glycome analysis based on MS. To improve the in-depth glycome analysis, it is also often necessary to separate the glycans (especially for glycan isomers characterization).

To isolate and purify glycans from other substances, it is necessary to first understand the physical and chemical properties of glycans very well to design and develop appropriate methods. In fact, the isolation and enrichment methods of N-glycans currently used can be divided into two types, direct affinity and chemical interaction based strategies. The former mainly includes porous graphitic carbon (PGC), hydrophilic interaction liquid chromatography (HILIC), lectin-affinity (Kaji et al. 2003), which are common enrichment strategies widely used in modern MS-based glycomic experiments. The latter mainly utilize chemical reaction to introduce the separation and enrichment tag on the reactive groups of N-glycans (such as reducing end, carboxyl group on sialic acid residues). While the specificity of lectins for recognizing different types of glycans only facilitates the isolation of a subset of glycan from a pool, HILIC and PGC-SPE can be employed for isolation of a broad range of glycans, making the latter two adsorption chromatography methods applicable in a wide range of glycomics studies (Selman et al. 2011). Thus, we will emphasize on addressing HILIC and PGC approaches.

3.2.2.1 Porous Graphitic Carbon (PGC) Enrichment and Separation

PGC is a popular stationary phase material for N-glycans purification and separation. The structure of PGC is formed by winding irregular two-dimensional graphite ribbons. PGC resin captures glycans through hydrophobic interactions and polar or electrostatic interactions, which are thought to be due not only to the presence of oxygen-containing functional groups, but also to a more complex set of interactions that are not fully understood (Stine 2017).

PGC is commonly incorporated into solid phase extraction (SPE) cartridge or 96-well plates for the purification of N-glycans released from glycoprotein(s) or complex biological samples prior to MS analysis.

Meanwhile, owing to the value of the mixed mode of retention on PGC material, PGC is not only used for purification and enrichment of glycans, but also becomes a significant chromatographic material for separation of glycans, with good resolution for isomers.

There are a large number of studies employing PGC as a SPE cartridge or chromatographic stationary phase since they have been invented more than 30 years (Gilbert et al. 1982; Hounsell 1994). A recent study by Seo and coworkers demonstrated great performance of PGC in purifying and fractionating glycans with different chemical properties due to its unique binding/elution behaviour (Seo et al. 2019). In their work, heterogeneous glycans including neutral and multiple-acidic glycans released from a therapeutic lysosomal enzyme, agalsidase-beta, were selectively purified and fractionated with PGC-SPE according to the size and polarity of glycans, followed by LC-MS/MS analysis. With this method, a total of 52 glycans containing 18 neutral, 19 phosphorylated and 15 sialylated glycans were successfully characterized. It showed that in-depth analysis of heterogeneous glycans can be realized by using stepwise SPE glycocapture in conjugating with simple one-dimension LC/MS.

One major of advantages of PGC chromatography is its powerful separation capability in discrimination between glycan isomers, enabling in-depth N-glycome analysis. It has been shown successfully applied to separate isomers of unlabelled oligomannosidic glycans (Pabst et al. 2012), sialylated N-glycans (Pabst et al. 2007; Stadlmann et al. 2008), as well as fucosylated N-glycans (Chen et al. 2020; Everest-Dass et al. 2012). In depth N-glycomic profiling from different glycan samples, such as glycans from dry blood spots (Ruhaak et al. 2012) and blood serum (Aldredge et al. 2012; Song et al. 2015) were reported by Lebrilla and coworkers, by utilizing PGC chip coupled with MS. It was shown that more than 170 N-glycan structures were separated and profiled based on first fractionation by PGC-SPE, then separation and analysis by LC chip with PGC column coupled to quadrupole time-of-flight (Q-TOF) MS (Song et al. 2015).

It was also found that optimization of separation temperature is important for highly efficient separation of glycans. Zhou et al. reported that baseline resolution of linkage isomers and monosaccharide positional isomers for both neutral and sialylated permethylated glycans were achieved on a PGC column at elevated temperatures of 75°C (Figure 3.2). Thanks to enhancing separation efficiency, a total of 127 unique glycan structures with 39 isobaric structures including 106 isomers and 21 nonisomeric

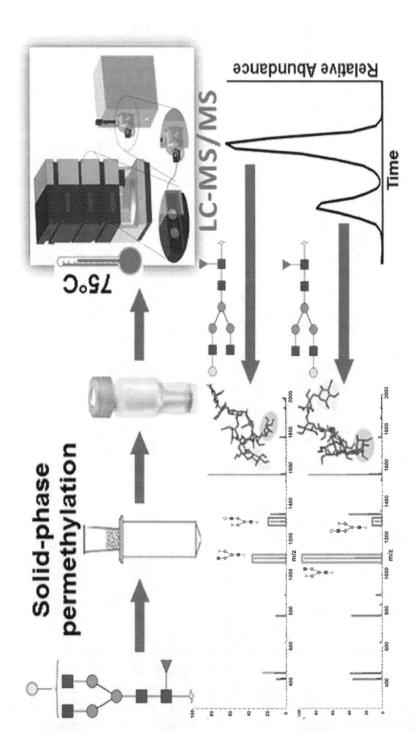

FIGURE 3.2   A pipeline for isomeric N-glycan identification based on LC-MS-MS. Powerful separation capability in discrimination between permethylated glycan isomers can be achieved with PGC chromatography under optimized temperature, enabling in-depth N-glycome analysis. (Reprinted from Zhou et al. (2017b) with permission from American Chemical Society.)

glycans were identified by PGC-LC-MS/MS in their work (Zhou et al. 2017b). Enhancement of fucosylated native N-glycan isomer separation was also demonstrated with an ultrahigh column temperature at 190°C in PGC-MS more recently (Chen et al. 2020).

In practice, it is often necessary to optimize the method according to the characteristics of interfering substances introduced in the sample pretreatment process, and sometimes two different purification methods need to be combined to achieve highly efficient purification and enrichment. For example, Wang and coworkers have quantified isomer-specific sialylated N-glycans released from whey glycoproteins in human milk. In their work, C18 SPE column was first used for removal of proteins and peptides in the sample solution and then a PGC-SPE N-glycans was utilized to desalt and enrich N-glycans in the elution from C18 tip (Jin et al. 2021). Manfred and coworkers reported a high-throughput sample preparation method based on 96-well platform for using PGC-LC-MS. The released and reduced N-glycans and O-glycans are first desalted by cation exchange column, and then further purified and enriched by PGC in their method (Zhang et al. 2020b).

Similarly, the orthogonal selectivity between PGC and other stationary phases can be effectively exploited in two-dimensional liquid chromatographic (2D-LC) systems to obtain high efficiency and resolution for separation of N-glycans, such as with HILIC. Liang and coworkers showed that a total of 31 N-glycan compounds including 7 pairs of isomers, most of them with the purity higher than 90%, were successfully isolated from ovalbumin (OVA) based on two-dimensional HILIC-PGC system (Cao et al. 2021).

Here are just a few examples of method development and applications based on PGC. For more detailed examples of related studies, readers may refer to some review literatures (Stine 2017; Zhang et al. 2018).

3.2.2.2 Hydrophilic Interaction Chromatography (HILIC) Separation

HILIC is a commonly used tool for enrichment and separation of polar and hydrophilic compounds including N-glycans, which is able to retain both native and derivatized N-glycans. Like PGC, HILIC can also be utilized both as an enrichment mode prior to MS analysis of N-glycans and as the chromatography mode used in liquid chromatography directly coupled to MS analysis. HILIC employs highly polar (hydrophilic) stationary phases, typically such as bare silica, amide-, amino-, hydroxyl-bonded, or zwitterionic on silica and polymeric supports (Zauner et al. 2011). Glycans prone to being retained on the stationary phase in the presence of mobile phase with a high content of organic solvent (mostly acetonitrile) due to promoted hydrophilic interactions between the analyte and the polar stationary and then be eluted with an increasing aqueous gradient (Zauner et al. 2011). The retention time of the N-glycans on the HILIC column is determined by their hydrophilic potential and depends on the structure properties of the glycan such as size, composition, linkage and oligosaccharide branching. As a result, many glycan structural isomers can be separated using HILIC prior to a MS analysis (Isokawa et al. 2014; Tao et al. 2014; Veillon et al. 2017a).

HILIC SPE microtips based on different materials, such as sepharose, microcrystalline cellulose and cotton, are used for microscale purification and enrichment of N-glycans (Krishnamoorthy and Mahal 2009; Selman et al. 2011). Cotton HILIC approaches have attracted more interests due to advantages of simplicity, high selectivity, economy, universality and broad applicability for isolating both native and derivative glycans from biological samples prior to MS analysis. It was shown that HILIC SPE microtips packed with cotton wool composed of virtually pure cellulose can efficiently purify and enrich both neutral and acidic N-glycans released from IgG and transferrin followed by MS detection (Selman et al. 2011). Peng et al. reported that the enrichment selectivity towards N-glycan was improved approximately 100-folds than conventional methods when using sterilized cotton as a SPE matrix for selective N-glycan enrichment (Peng et al. 2019a). Han et al. presented a new method for the purification of N/O-glycans and their derivatives from glycoprotein using a hand-packed absorbent cotton HILIC. They demonstrated high binding capacity and good recovery of glycans with this method and successful application to the analysis of human serum and foetal bovine serum glycoprotein N-glycans.

The early application of HILIC in separation of native and reduced N-glycans has been summarized in detail previously elsewhere, such as for large-scale total N-glycome profiling study and fast high-resolution separation in ultra-performance liquid chromatography (UPLC) form coupled with MS (Zauner et al. 2011). In recent years, many studies have developed new methods to improve the ability to analyze glycan isomers using HILIC-MS in low abundance, or with highly sialylated glycans isomers as well as analysis speed. For example, HILIC has limitation in separation of large polar molecules such as highly sialylated glycans. The reason is HILIC chromatographic efficiency drops because dispersion of the aqueous partition layer surrounding the stationary phase particles increased as the aqueous gradient increases, resulting in broad peaks of highly retained complex-glycans and poor capability for separation of highly sialylated N-glycan isomers (Tousi et al. 2013). Tousi et al. reported that they realized linkage specific characterization of highly sialylated glycans by coupling 4-(4,6-dimethoxy-1,3,5- triazin-2yl)-4-methylmorpholinium chloride (DMT-MM) charge neutralization of sialic acids with nano-HILIC-Orbitrap-MS. DMT-MM derivatization can improve the preceding HILIC separation prior to MS by increasing the hydrophobicity and altering the selectivity of the glycans (Tousi et al. 2013). Tao et al. developed a method to improve separation of sialylated N-Glycans linkage isomers by employing a novel superficially porous particle (Fused-Core) Penta-HILIC (hydrophilic interaction liquid chromatography) column (Tao et al. 2014). Similar to PGC, HILIC is also used in combination with other chromatographic column to achieve high resolution of glycan isolation (Cao et al. 2021).

3.2.2.3 Enrichment Methods Based on Chemical Reaction
The major drawback of the above-mentioned affinity-based approaches is nonspecific binding. As a result, some enrichment methods based on chemical reaction were developed to improve specificity in enrichment. They utilized different functional groups in glycans, such as aldehyde group on reducing end, carboxylic acid on sialic acid and uronic acid, amine on N-acetylhexosamine to form cleavable bond on solid phase or introduce a tag which can further form reversable binding onto solid phase.

Many different works have utilized solid phase with functional groups such as hydrazide, hydroxylamine, cysteinamide and amine to capture and release native glycans by direct formation of cleavable bond of hydrazone, oxime, thiazolidine and imine, respectively in different studies (Larsen et al. 2006; Yang and Zhang 2012; Zatsepin et al. 2002). The bond can be formed and cleaved under different conditions. Such as glycans can be conjugated to hydrazide groups on solid materials and recovered by heating under acidic condition after removal of non-specific binding components (Furukawa et al. 2008). The major methods of glycan isolation based on forming cleavable bond on solid-phase materials have been summarized in literatures, where more examples of these methods and their advantages and disadvantages are addressed (Larsen et al. 2006; Yang and Zhang 2012).

In addition to the aforementioned method that native glycan is directly captured by the solid phase material by chemical reaction, another enrichment strategy based on solid-phase extraction is introduction of enrichment tags by chemically labelling at the reducing end in N-glycans. As a result, the specific enrichment of glycans were achieved through the specific binding of enrichment tag and solid materials. Lu group have developed such methods with different enrichment tags, such as perfluoroalkyl tag (Li et al. 2015), 4-aminophenylphophate (Zhang et al. 2016), amino acid coded affinity tag (Wang et al. 2019). They demonstrated that the introduced tags can not only largely enhance specific enrichment of N-glycans, but also improve the ionization efficiency of glycan and discrimination of glycan isomers due to improved signals of fragments in tandem spectra. Another advantage of tag-assisted methods shown was easy introduction of isotopic labelling on tag group, achieving accurate and reproducible quantification of N-glycans (Figure 3.3) (Yang et al. 2019).

3.2.3 Derivatization of N-Glycans

Although MS has now become an indispensable tool for glycan analysis, there are some challenges in analysis of glycans by MS due to some intrinsic characteristics of glycan, such as low ionization efficiency, the labile nature of glycan residues (mostly sialic acid), complicated branched structures and various linkage isomers (Nishikaze 2017a). Chemical derivatizations of N-glycans prior to MS analysis are often carried out to overcome these challenges. There are three major types of modification, including permethylation of their hydroxyl- and the N-acetyl- and carboxylic acid-groups, labelling of reducing ends and sialic acid derivatization. Among them, permethylation and reducing end labelling can be carried out for both neutral and acidic N-glycans. Figure 3.4 provides commonly used derivatization approaches of permethylation and reducing end labelling (Veillon et al. 2017b).

3.2.3.1 Permethylation

Permethylation of N-glycans has been commonly and successfully applied in various structural analysis of glycans by MS, which is classically carried out in dimethyl sulfoxide (DMSO) with sodium hydroxide as base and

Sample A

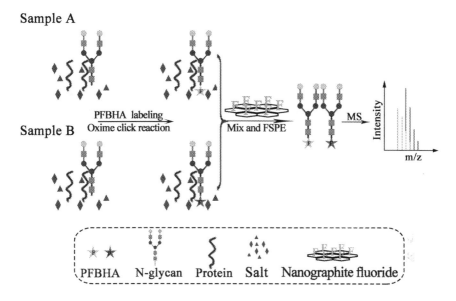

FIGURE 3.3 An integrated pipeline that combines isotopically fluorous tag label-ling and fluorous solid-phase extraction to quantitatively analyze the N-glycome by MS. Light and heavy aminoxy-functionalized fluorous tags (PFBHA and PFBHA-d2) were used to label the N-glycans through the oxime click reaction. Then the fluorous tag labelled N-glycan could be purified through the fluorous solid-phase extraction from contaminants like salts and proteins for the follow-ing quantitative analysis by MS. (Reprinted from Yang et al. (2019) with permis-sion from American Chemical Society.)

iodomethane as methylating reagent. With permethylation, all hydrogens linked to oxygen and nitrogen atoms are theoretically replaced by methyl groups. Permethylation provides several following major advantages. The hydrophobicity of the glycans can be greatly enhanced which provides improved glycan ionization efficiency in MS and chromatographic sepa-rations. The stability of sialic acid residue can be enhanced, reducing loss of sialic acid during MS analysis. As a result, entire N-glycome including both neutral and acid glycans can be profiled in the MS positive-ion mode. Cross-ring fragmentation during tandem MS analyses can be enhanced, providing more definitive structural glycan information such as sequence, branching and linkage. Also, it can prevent intramolecular migration of fucose units during MS or tandem MS analysis because all active sites pos-sible for fucose migration were occupied by methyl group (Mucha et al.

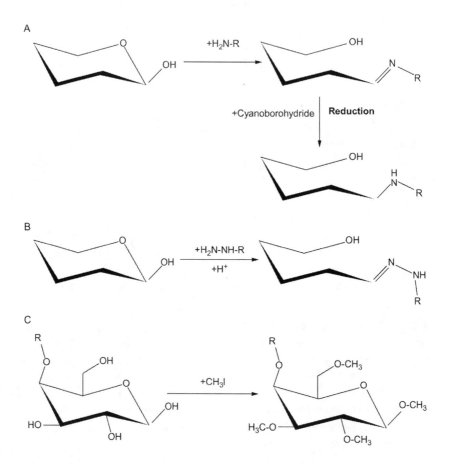

FIGURE 3.4 Commonly used derivatization approaches by reducing end label-ling including reductive amination (A) and hydrazide derivatization (B); and permethylation (C). (Reprinted from Veillon et al. (2017b) with permission from Elsevier.)

2018; Zhou et al. 2017a). Moreover, introducing stable isotope labelling by using heavy-labelled methyl iodide enables relative and absolute quantita-tion of glycan. Precise quantitative N-glyome comparison between dif-ferent blood samples were demonstrated in a single MALDI-MS analysis (Kang et al. 2007).

Traditional protocols for permethylation tend to be labour inten-sive, requiring manual permethylation and liquid-liquid extraction steps (Shubhakar et al. 2016). Recent advances on glycan permethyl-ation mainly focused on making the method with less side-product,

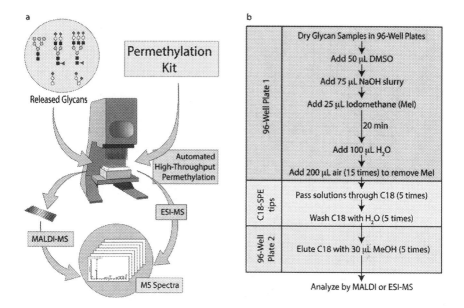

FIGURE 3.5 (A) Schematic showing permethylation procedure carried out with an automatic 96-well pipet. (B) Flowchart showing the steps involved in the simplified automated permethylation workflow. The permethylated glycans are passed from Plate 1 to the C18 SPE tips, and from there to Plate 2. (Reprinted from Shajahan et al. (2019) with permission from American Chemical Society.)

simpler, faster, automated, high-throughput and suitable for labelling low amounts of samples. For example, many efforts such as introducing solid phase micro spin columns, 96-well plate format, a liquid handling robot were attempted to achieve above-mentioned aims (Shajahan et al. 2019; Shubhakar et al. 2016). Figure 3.5 showed one example protocol (Shajahan et al. 2019).

3.2.3.2 Derivatization on Reducing Ends
Derivatization on reducing ends usually through reductive amination or hydrazide is another type of the most popular derivatizations except permethylation. There are several benefits possibly introduced by derivatization at reducing ends of glycans, such as providing N-glycan with fluorescent label, improving chromatography separation and ionization efficiency in MS due to enhanced hydrophobicity of N-glycan, assisting reliable glycan structure identification and quantitation (Li et al. 2015).

A variety of reducing ends labelling reagents have been employed for MS-based glycomics, such as 2-aminobenzoic acid (2-AA) (Anumula and Dhume 1998), 2-aminobenzamide (2-AB) (Royle et al. 2008), procainamide (ProA) (Klapoetke et al. 2010), RapiFluor-MS (or RFMS) (Lauber et al. 2015), Carboxymethyl trimethylammonium hydrazide (Girard's T reagent) (Gil et al. 2008). The techniques of glycan derivatization by reductive amination and hydrazide labelling with these reagents have been overviewed in some reviews (Zhang et al. 2018; Zhou et al. 2017c). Since each of derivatization methods provides different properties associated with performance in both glycan retention on LC columns and MS analyses. Zhou et al. have performed a comprehensive assessment of 2-AB, ProA, aminoxyTMT, RFMS labelling, reduction and reduction with permethylation for N-glycan analysis based on LC-MS. The results demonstrated that RFMS provided the highest MS signal enhancement for neutral glycans, while permethylation significantly enhanced the MS intensity and structural stability of sialylated glycans among the derivatization strategies examined (Zhou et al. 2017c).

Recently, in order to reduce the sample processing time and sample loss while improving the ionization efficiency of N-glycan analysis based on MALDI-MS, some on-target-based derivatization methods were proposed, such as 3-aminoquinoline (Rohmer et al. 2010), aniline (Snovida et al. 2006), Girard's T, 2-AA (Hronowski et al. 2020). These methods are performed directly on the MALDI target plate without removing the derived reagents, reducing time and sample loss while enhancing ionization efficiency. Recent advances in on-target derivatization methods have been focused on the development of reagents that can be used both as labelling reagents and as co-matrix, as well as for the production of homogeneous sample to facilitate quantification. Lu group has employed aminopyrazine (Cai et al. 2013) and hydrazinonicotinic acid (HYNIC) (Jiao et al. 2015) as the derivatization reagent as well as the comatrix to enhance the N-glycan ionization in MALDI-TOF MS. Zhao et al. employed a novel reactive matrix, 2,5-dihydroxybenzohydrazide (DHBH) to form uniform cocrystallization of analytes-matrix mixtures with catalytic matrix DHB (Zhao et al. 2019). Another representative example is that 2-AA was recently employed as on a MALDI target N-glycans labelling reagents through nonreductive amination and simultaneously as a matrix in MALDI-MS glycan analysis (Hronowski et al. 2020). All these methods demonstrated that removal of the excess labelling reagents was not necessary as they can be simultaneously used as MALDI-MS matrix, reducing sample loss and sample pretreatment time.

3.2.3.3 Sialic Acid Derivatization

Sialic acid (SA) residues are usually linked to non-reducing ends of N-glycans with different linkages (mainly α2,3-/α2,6-linkages). The presence of strong negative charges on sialic acids led to their low ionization efficiency and instability in the MS-based analysis. Thus, it is difficult to profile entire glycome (including neutral and acidic glycans) and quantitate in a single MS analysis as well as in a high-throughput manner without SA protection.

Many chemical derivatization methods were developed to neutralize and stabilize these sialic acids prior to MS analysis. Besides permethylation, esterification and amidation of the carboxyl group in sialic acid also provided effective solutions to protect the sialic acids. At first, different reagents were used to form ester or amide with carboxyl groups on sialic acids, such as methyl iodide to methylester (Powell and Harvey 1996), ammonium chloride (NH_4Cl) with DMT-MM as a condensing reagent to amide (Sekiya et al. 2005), methylamine chloride with powerful condensing reagent of PyAOP to methylamide (Liu et al. 2010), normal and heavy (D9) p-toluidine for amination (Shah et al. 2013). All of above-mentioned methods are linkage-nonspecific sialic acid derivatization.

Considering α2,3- and α2,6-linkage of glycans can be associated with different biological processes (Suzuki 2005; Suzuki 2006; Zhang et al. 2019), recent advances on sialic acid derivatization mainly focused on developing highly efficient linkage-specific approaches. The principle behind the currently reported strategies is taking advantage of the different reactivity of 2,3- and 2,6-sialic acid and producing products with different mass for 2,3- and 2,6-sialic acid containing glycans after derivatization. Under strictly controlled reaction conditions, 2,6-linked sialic acids react with external nucleophiles to form esterification and amidation labelling, whereas 2,3-linked sialic acids form lactones by intramolecular dehydration (Nishikaze 2019). Originally, same nucleophile and condensing reagents but different reaction conditions with linkage-nonspecific derivatization as mentioned above were employed for linkage-specific modification, such as utilizing MeOH (nucleophile) and DMT-MM (condensing reagent) (Wheeler et al. 2009), ammonia (nucleophile) and DMT-MM (Alley and Novotny 2010). Later, a new method employing EDC/HOBt as condensing reagent mixture and ethanol as a nucleophile was reported, resulting in esterification and lactonization for 2,6- linked and 2,3-linked sialic acids respectively (Reiding et al. 2014). This approach appears to be robust and practical because the derivatization reaction can be performed directly in PNGase F released N-glycan mixture without a prior desalting, facilitating automated and large scale glycomics analysis of plasma N-glycans (Vreeker et al. 2018). However, the derivatization reaction of these methods have to be performed just before MS analysis because the lactone forms has been found unstable after formation (Wheeler et al. 2009).

To address the issues of lactone instability, some new methods were further developed in recent years, where one more step reaction after above-mentioned derivatization step was introduced to convert the lactone forms into more stable forms (Nishikaze 2019). Permethylation, lactone cleavage followed by amidation and direct aminolysis are some reported approaches for stabilization of lactone form. For example, Jiang et al. combined linkage-specific dimethylamidation with an in-solution permethylation technique to produce dimethylamidated and methylesterified forms for $\alpha2,6$- and $\alpha2,3$-linked sialic acids respectively (Jiang et al. 2017). Nishikaze et al. employed sequential two-step alkylamidations to selectively produce amidated $\alpha2,6$- and $\alpha2,3$-linked sialic acids with different length of alkyl chain, enabling distinction of $\alpha2,3$-/$\alpha2,6$-linkage isomers from given mass spectra, as shown in Figure 3.6 (Nishikaze et al. 2017b). The method for quantitation of sialyl-linkage isomers was further developed by combining stable-labelling with sialic acid linkage-specific derivatization (Peng et al. 2019b). More detailed examples and discussion about sialic acid derivatization for MS analysis can be referred to several reviews (Nishikaze 2019; de Haan et al. 2020).

3.3 QUALITATIVE N-GLYCOMICS METHODS

Comprehensive characterization of N-glycome structures is particularly challenging due to the structural complexity and their isomers. MALDI-MS and ESI-MS are two commonly used MS techniques for N-glycome characterization because they provide a link between mass and composition and the capacity of performing tandem MS (MS^2 and MS^n). N-glycan structural elucidation based on MS mainly depends on accurate mass weight provided from high resolution mass spectrometry (HRMS) and fragment ions produced from tandem MS analysis. As we discussed

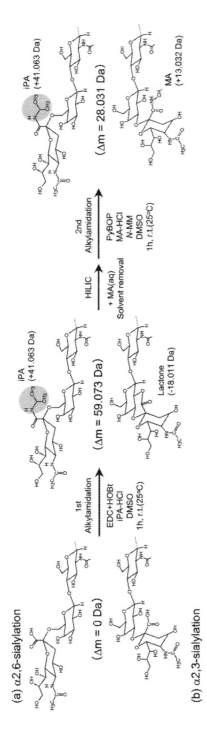

FIGURE 3.6 Reaction scheme of sialic acid linkage specific alkylamidation (SALSA). Derivatization of (A) α2,6-linked and (B) α2,3-linked Neu5Ac residues on the terminal Gal-GlcNAc are depicted. SALSA employs two sequential reactions, isopropylamidation, and methylamidation. In the first alkylamidation, α2,6- and α2,3-linked sialic acids are selectively converted to their iPA-derivatized and lactone forms, respectively. After the removal of excess reagents by HILIC SPE, MA(aq) is added to promote alkali-driven lactone cleavage, and solvents were then removed in vacuo. In the second alkylamidation, lactonized and delactonized α2,3-linked sialic acids are derivatized by MA under mild reaction conditions. (Reprinted from Nishikaze et al. (2017b) with permission from American Chemical Society.)

in the first part, chromatographic or electrophoretic separations and glycan derivatization techniques are often employed prior to MS analysis to improve separation of glycans and their isomers to achieve in-depth glycome analysis. Thus, by combining with optimized sample pretreatment conditions including all or some of key steps as mentioned in the aforementioned sample preparation, MALDI or ESI-MS (MS^n) analysis have been successfully used for both profiling of N-glycome from complex biological sample and characterization of the glycan structures (North et al. 2010; Pang et al. 2011).

For example, detailed structure with linkage positions was characterized according to the diagnostic fragment ions from the cleavages of glycosidic bonds and cross-ring cleavages of sodium adducts of permethylated glycan in the tandem MS data by using high-energy collision-induced dissociation (CID) MALDI-TOF/TOF-MS (Mechref et al. 2003). MALDI-MS is an effective technique for N-glycome analysis due to its high analysis speed, low sample expenditure, high tolerance of salt, high-throughput capability and high automation potential (Liang et al. 2019; Yamagaki 1999). Automated and high-throughput analysis of permethylated serum N-glycome was demonstrated by using MALDI linear ion trap mass spectrometry with Glycan structures confirmed by MS^2 analysis (Guillard et al. 2009). More recently, Reiding et al. demonstrated that MALDI-TOF-MS enabled high-throughput N-glycome analysis and provided compositional information on high complex N-glycans. They also showed that MALDI-TOF-MS could identify the linkage-specific sialic acid isomers by employing linkage-specific sialic acid derivatization (Reiding et al. 2019). MALDI-MS based N-glycomic analysis has been extensively applied to discover glycan biomarkers of diseases where high throughput and high speed are usually needed (Sun et al. 2019).

Except for the structure identification of N-glycome directly by MS and tandem MS without separation, LC-MS or CE-MS coupling were also very often employed to perform N-glycome analysis. Retention and migration times, precursor masses and fragmentation spectra may then be used for glycan structural elucidation (Pabst et al. 2007). Hinneburg et al. demonstrated that Nano PGC LC-MS combined with negative-mode tandem mass spectrometry allowed the structural elucidation of unknown glycans in N-glycome and O-glycome from minor amount of tissues (Hinneburg et al. 2017). The relative abundances of individual glycans along with detailed isomer and functional epitope structural information of N- and

O-glycans can be obtained with this technique (Hinneburg et al. 2017). Recently, Wuhrer group presented a novel method for the highly sensitive and in-depth characterization of released N-glycans by combining techniques of capillary electrophoresis separation, electrospray ionization-mass spectrometry, linkage-specific derivatization of sialic acids and uniform cationic reducing end labelling of all glycans (Lageveen-Kammeijer et al. 2019). In their method, linkage-specific sialic acid neutralization reaction was performed to neutralize all glycans and enable the linkage-specific sialic acids differentiation. Permanent cationic labelling of the N-glycan reducing ends with hydrazide Girard's reagent P (GirP) was introduced to enable electrophoretic migration of the neutralized N-glycans and enhance ionization efficiency in positive-ionization mode. Accurate mass and isotopic pattern were used to assign N-glycan compositions and distinct diagnostic ions produced by tandem MS via CID were used to characterize and confirm glycan structures (Lageveen-Kammeijer et al. 2019).

To obtain in-depth compositional and structural information, one single method seems insufficient and multimethodological approach was proposed. Several complementary glycan profiling techniques have been combined to identify N-glycans released from the glycoproteins in human urinary exosomes (Song et al. 2019). In their work, CE-MS, MALDI-MS and LC-MS/MS were used to identify N-glycan and sulfated N-glycan compositions. Meanwhile, sialidase treatment and sialic-acid linkage specific alkylamidation followed by MALDI-MS analysis were employed to assign sialyl-linkage isomers. As a result, this multimethodological approach enabled identification of 219 N-glycan structures consisting of 175 compositions, 27 sulfated glycans, 26 structural isomers and 64 sialic-acid linkage isomers.

One crucial technique of MS-based glycomics analysis is tandem MS (MS^n) which generated fragment ions for detailed glycan structural characterization. It was notable that some other dissociation techniques besides CID for glycomics have been developed, attempting to provide more fragment ion information over typically used CID-based MS^n method, such as electronic excitation dissociation (EED) (Wei et al. 2020) and ultraviolet photodissociation (UVPD) (Morrison and Clowers 2017). In contrast to CID which is low-energy dissociation technique not capable of providing sufficient fragmentation information, EED and UVPD are high-energy dissociation techniques which appear to provide extensive fragmentation information, in particular cross-ring fragments that can distinguish glycan isomers. However, because of complex fragmentations generated by EED and UVPD, more efficient methods for interpretation of mass spectra to determine the glycan structures are expected to be explored.

3.4 METHODS FOR QUANTITATIVE N-GLYCOMICS ANALYSIS

Practical N-glycomics quantitative methods remain urgently needed in the fields of disease marker discovery, protein glycosylation function research and quality control of biopharmaceuticals. Currently, the methods of quantitative N-glycomics analysis based on MS mainly include label-free, isotopic labelling and internal standard normalization strategies.

For label-free method, the relative abundance of each targeted glycan is calculated by dividing abundance of each glycan by total abundances of all glycans. Comparative quantitation of each glycan was then obtained by parallel analysis of all the samples across multiple MS analyzes. (Bladergroen et al. 2015; Gil et al. 2010; Kang et al. 2008; Toyoda et al. 2008). This method has the advantages of simple experimental procedure and low experimental cost. However, it might be affected by matrix effect and ionization fluctuation of MS and other factors to induce quantitative deviation among different runs.

Stable isotope labelling quantification can introduce stable isotope labelling at reductive and/or non-reductive ends of glycans by metabolic labelling (Orlando et al. 2009), enzymatic labelling (Cao et al. 2015; Yang et al. 2013; Zhang et al. 2011), and chemical labelling (Dong et al. 2019; Feng et al. 2019; Zhang et al. 2015; Zhou et al. 2016a). Then, the samples labelled with light and heavy isotopes were mixed and analyzed simultaneously based on MS to achieve comparative glycan quantitation, which might reduce errors induced by the fluctuation of glycan ionization efficiency and instrument response across different runs. Metabolic labelling is the introduction of isotopic labelling through cell metabolism, which is generally only applicable to living systems. Enzymatic labelling is the introduction of isotope labelling through $H_2{}^{18}O$ in the process of enzymatic release of N-glycans. However, the high cost of ^{18}O-water might limit the methods to small samples (Schwartz et al. 2016).

Chemical labelling is the most common stable isotopic labelling method, where stable isotopes can be incorporated into enzymatic released N-glycans via chemical reaction. Labelling tags with stable isotopes of ^{2}D, ^{13}C, or ^{15}N are often introduced through a varies of chemical reactions such as permethylation, amination or hydrazide. For example, 2-plex tags by using aminobenzoic acid (2-AA) isotope-coded reagent, such as (d0) 2-AA and (d4) 2-AA (Bowman and Zaia 2007), 2-12[C6]-AA and 2-13[C6]-AA (Martin et al. 2015), were used as mass tags to label the glycan reducing end based on the reaction at its terminal aldehyde group. Similarly, permethylation, another common derivatization method, was employed to develop 2-plex tags with different isotope-coded reagents, such as CH_3I/CD_3I (Kang et al. 2007) and $^{12}CH_3I/^{13}CH_3I$ (Alvarez-Manilla et al. 2007). As the 2-plex isotope labelling methods are only capable of simultaneously analyzing two samples. Multiplex isotope tags were also introduced to achieve comparative glycan quantitation, such as using CH_2DI, CHD_2I, CD_3I, $^{13}CH_3I$, $^{13}CH_2DI$, $^{13}CHD_2I$, $^{13}CD_3I$ and CH_3I as permethylation reagents to attain 8-plex sample analysis at a time (Dong et al. 2019). In addition, isobaric labelling methods were also introduced for quantitative glycomics, which have high multiplexing capacity by tandem MS, such as isobaric aldehyde reactive tags (iARTs) (Yang et al. 2013), GlycoTMT (Hahne et al. 2012) and AminoxyTMT (Afiuni-Zadeh et al. 2016). Recent research efforts on isobaric labelling methods seems focused on improving the yield of report ions, reducing the cost of the labelling reagents and increasing sample throughput (Feng et al. 2019; Zhou et al. 2016a). These chemical labelling quantification methods usually require chemical labelling in each of sample. Thus, the experimental process tends to be tedious and the experimental cost is high, which is usually only used for quantitative analysis of a limited number of samples. In addition, special attention should be paid to avoid sample loss during additional labelling and purification steps (Qin et al. 2019).

Some simple quantitative glycomics methods have been developed by spiking into the sample with exogenous glycans as internal standards that can reduce the deviation caused by each parallel operation. For example, malto-series oligosaccharides (Mehta et al. 2016) and isotopic-labelled N-glycans from purified ^{15}N labelled monoclonal antibody (Zhou et al. 2016b) have been demonstrated effective as an internal standard to improve the reliability of quantitative glycomics. However, as these glycan internal standards lack diverse glycan coverages, it remains a challenge to realize the effective normalization of each of all diverse glycans. More recently, Qin et al. presented a new simple and economic method to provide a complete set of glycan internal standards named as "Bionic

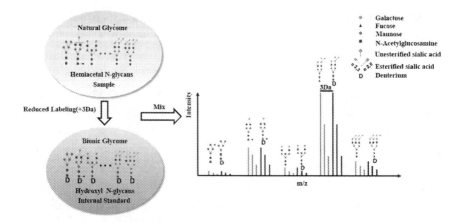

FIGURE 3.7 The strategy for N-glycome quantitation using bionic glycome internal standard based on MALDI-MS. Bionic glycome was prepared through one step reducing reaction with $NaBD_4$ using nature N-glycome to be analyzed as substrate. Each bionic glycan has 3 Da mass increment over its corresponding glycan analyte due to hemiacetals/alditols and H/D mass difference. Bionic glycome has the same glycome composition and similar glycome profile in abundance with N-glycome to be analyzed from biological sample. (Reprinted from Qin et al. (2019) with permission from Elsevier.)

Glycome" for quantitative N-glycomic analysis (Qin et al. 2019), as shown in Figure 3.7. "Bionic Glycome" is obtained by pooling all the samples to be analyzed. This pool is then digested with PNGase F, reduced and simultaneously labelled with $NaBD_4$ and subsequently esterified. Samples to be analyzed quantitatively are treated in the same way, skipping the reduction step with $NaBD_4$. As a result, the N-glycans to be quantified have a molecular weight of 3 units less than their corresponding bionic glycans due to hemiacetals/alditols and H/D mass difference. And Bionic Glycome provide a N-glycan internal standard library with good coverage of the natural N-glycome to be analyzed in the biological sample as they have the same glycome composition and similar glycome profile in abundance. Reliable quantitation of the N-glycome of proteins from simple or complex biological system was demonstrated in this method. The Bionic Glycome has the same glycome composition and similar glycome profile in abundance with N-glycome to be analyzed from biological sample. The method provided a facile general way to obtain internal standard with good coverage for all glycan species of interest for easy and accurate N-glycome quantitation, which showed great potential in discovery of glycan biomarker in large-size samples (Qin et al. 2019).

3.5 CONCLUSIONS

As widely described in this chapter, great progress has been made in MS-based glycomics analysis, providing a powerful analytical tool for both qualitative and quantitative characterization of N-glycans. Both progress on MS technology and sample preparation techniques are very important and remains demanded. Another leap in analytical capability could benefit from advances in these key technologies, further advancing glycan biomarker discovery and a better understanding of the structure-function relationship of glycomics.

REFERENCES

Afiuni-Zadeh, S., J. C. Rogers, S. I. Snovida, R. D. Bomgarden, T. J. Griffin. 2016. AminoxyTMT: A novel multi-functional reagent for characterization of protein carbonylation. *Biotechniques* 60: 186–8, 190, 192–6.

Aldredge, D., H. J. An, N. Tang, K. Waddell, C. B. Lebrilla. 2012. Annotation of a serum N-glycan library for rapid identification of structures. *Journal of Proteome Research* 11: 1958–68.

Alley, W. R., M. V. Novotny. 2010. Glycomic analysis of sialic acid linkages in glycans derived from blood serum glycoproteins. *Journal of Proteome Research* 9: 3062–72.

Altmann, F., S. Schweiszer, C. Weber. 1995. Kinetic comparison of peptide N-glycosidase-F and N-glycosidase-a reveals several differences in substrate-specificity. *Glycoconjugate Journal* 12: 84–93.

Alvarez-Manilla, G., N. L. Warren, T. Abney, et al. 2007. Tools for glycomics: Relative quantitation of glycans by isotopic permethylation using (CH3I)-C-13. *Glycobiology* 17: 677–87.

Anumula, K. R., S. T. Dhume. 1998. High resolution and high sensitivity methods for oligosaccharide mapping and characterization by normal phase high performance liquid chromatography following derivatization with highly fluorescent anthranilic acid. *Glycobiology* 8: 685–94.

Bladergroen, M. R., K. R. Reiding, A. L. H. Ederveen, et al. 2015. Automation of high-throughput mass spectrometry-based plasma N-glycome analysis with linkage-specific sialic acid esterification. *Journal of Proteome Research* 14: 4080–6.

Bowman, M. J., J. Zaia. 2007. Tags for the stable isotopic labeling of carbohydrates and quantitative analysis by mass spectrometry. *Analytical Chemistry* 79: 5777–84.

Cai, Y., Y. Zhang, P. Yang, H. Lu. 2013. Improved analysis of oligosaccharides for matrix-assisted laser desorption/ionization time-of-flight mass spectrometry using aminopyrazine as a derivatization reagent and a co-matrix. *Analyst* 138: 6270–6.

Cao, W. Q., W. Zhang, J. M. Huang, B. Y. Jiang, L. J. Zhang, P. Y. Yang. 2015. Glycan reducing end dual isotopic labeling (GREDIL) for mass spectrometry-based quantitative N-glycomics. *Chemical Communications* 51: 13603–6.

Cao, C. Y., L. Yu, J. Y. Yan, D. M. Fu, J. L. Yuan, X. M. Liang. 2021. Purification of natural neutral N-glycans by using two-dimensional hydrophilic interaction liquid chromatography x porous graphitized carbon chromatography for glycan-microarray assay. *Talanta* 221: 121382.

Chen, C. H., Y. P. Lin, C. T. Ren, et al. 2020. Enhancement of fucosylated N-glycan isomer separation with an ultrahigh column temperature in porous graphitic carbon liquid chromatography-mass spectrometry. *Journal of Chromatography A* 1632: 461610.

Cummings, R. D., J. M. Pierce. 2014. The challenge and promise of glycomics. *Chemistry & Biology* 21: 1–15.

De Haan, N., S. Yang, J. Cipollo, M. Wuhrer. 2020. Glycomics studies using sialic acid derivatization and mass spectrometry. *Nature Reviews Chemistry* 4: 229–42.

Dong, X., W. J. Peng, C. Y. Yu, et al. 2019. 8-plex LC-MS/MS analysis of permethylated N-glycans achieved by using stable isotopic Iodomethane. *Analytical Chemistry* 91: 11794–802.

Everest-Dass, A. V., D. Y. Jin, M. Thaysen-Andersen, H. Nevalainen, D. Kolarich, N. H. Packer. 2012. Comparative structural analysis of the glycosylation of salivary and buccal cell proteins: innate protection against infection by *Candida albicans*. *Glycobiology* 22: 1465–79.

Feng, Y., B. M. Chen, Q. Y. Yu, et al. 2019. Isobaric multiplex labeling reagents for carbonyl-containing compound (SUGAR) tags: a probe for quantitative glycomic analysis. *Analytical Chemistry* 91: 3141–6.

Figl, R., F. Altmann. 2018. Reductive alkaline release of N-glycans generates a variety of unexpected, useful products. *Proteomics* 18: 1700330.

Fischler, D. A., R. Orlando. 2019. N-linked glycan release efficiency: A quantitative comparison between NaOCl and PNGase F release protocols. *Journal of Biomolecular Techniques* 30: 58–63.

Furukawa, J. I., Y. Shinohara, H. Kuramoto, et al. 2008. Comprehensive approach to structural and functional glycomics based on chemoselective glycoblotting and sequential tag conversion. *Analytical Chemistry* 80: 1094–101.

Gil, G. C., Y. G. Kim, B. G. Kim. 2008. A relative and absolute quantification of neutral N-linked oligosaccharides using modification with carboxymethyl trimethylammonium hydrazide and matrix-assisted laser desorption/ionization time-of-flight mass spectrometry. *Analytical Biochemistry* 379: 45–59.

Gil, G. C., B. Iliff, R. Cerny, W. H. Velander, K. E. Van Cottt. 2010. High through-put quantification of N-glycans using one-pot sialic acid modification and matrix assisted laser desorption ionization time-of-flight mass spectrometry. *Analytical Chemistry* 82: 6613–20.

Gilbert, M. T., J. H. Knox, B. Kaur. 1982. Porous glassy-carbon, a new columns packing material for gas-chromatography and high-performance liquid-chromatography. *Chromatographia* 16: 138–48.

Guillard, M., J. Gloerich, H. J. C. T. Wessels, E. Morava, R. A. Wevers, D. J. Lefeber. 2009. Automated measurement of permethylated serum N-glycans by maldi-linear ion trap mass spectrometry. *Molecular Genetics and Metabolism* 98: 41.

Guo, R. R., G. Comamala, H. H. Yang, et al. 2020. Discovery of highly active recombinant PNGase H(+) variants through the rational exploration of unstudied acidobacterial genomes. *Frontiers in Bioengineering and Biotechnology* 8: 741.

Hahne, H., P. Neubert, K. Kuhn, et al. 2012. Carbonyl-reactive tandem mass tags for the proteome-wide quantification of N-linked glycans. *Analytical Chemistry* 84: 3716–24.

Helenius, A., M. Aebi. 2001. Intracellular functions of N-linked glycans. *Science* 291: 2364–9.

Hinneburg, H., F. Schirmeister, P. Korac, D. Kolarich. 2017. N- and O-glycomics from minor amounts of formalin-fixed, paraffin-embedded tissue samples. *Methods in Molecular Biology* 1503: 131–45.

Hounsell, E. F. 1994. Characterization of the glycosylation status of proteins. *Molecular Biotechnology* 2: 45–60.

Hronowski, X. L., Y. Wang, Z. Sosic, R. Wei. 2020. On-MALDI-target N-glycan nonreductive amination by 2-aminobenzoic acid. *Analytical Chemistry* 92: 10252–6.

Isokawa, M., T. Kanamori, T. Funatsu, M. Tsunoda. 2014. Recent advances in hydrophilic interaction chromatography for quantitative analysis of endogenous and pharmaceutical compounds in plasma samples. *Bioanalysis* 6: 2421–39.

Jiang, K., H. Zhu, L. Li, et al. 2017. Sialic acid linkage-specific permethylation for improved profiling of protein glycosylation by MALDI-TOF MS. *Analytica Chimica Acta* 981: 53–61.

Jiao, J., L. Yang, Y. Zhang, H. Lu. 2015. Hydrazinonicotinic acid derivatization for selective ionization and improved glycan structure characterization by MALDI-MS. *Analyst* 140: 5475–80.

Jin, W., C. Li, M. Zou, et al. 2021. A preliminary study on isomer-specific quantification of sialylated N-glycans released from whey glycoproteins in human colostrum and mature milk using a glycoqueuing strategy. *Food Chemistry* 339: 127866.

Kaji, H., H. Saito, Y. Yamauchi, et al. 2003. Lectin affinity capture, isotope-coded tagging and mass spectrometry to identify N-linked glycoproteins. *Nature biotechnology* 21: 667–72.

Kang, P., Y. Mechref, Z. Kyselova, J. A. Goetz, M. V. Novotny. 2007. Comparative glycomic mapping through quantitative permethylation and stable-isotope labeling. *Analytical Chemistry* 79: 6064–73.

Kang, P., Y. Mechref, M. V. Novotny. 2008. High-throughput solid-phase permethylation of glycans prior to mass spectrometry. *Rapid Communications in Mass Spectrometry* 22: 721–34.

Klapoetke, S., J. Zhang, S. Becht, X. L. Gu, X. Y. Ding. 2010. The evaluation of a novel approach for the profiling and identification of N-linked glycan with a procainamide tag by HPLC with fluorescent and mass spectrometric detection. *Journal of Pharmaceutical and Biomedical Analysis* 53: 315–24.

Krishnamoorthy, L., L. K. Mahal. 2009. Glycomic analysis: An array of technologies. *Acs Chemical Biology* 4: 715–32.

Lageveen-Kammeijer, G. S. M., N. De Haan, P. Mohaupt, et al. 2019. Highly sensitive CE-ESI-MS analysis of N-glycans from complex biological samples. *Nature Communications* 10: 2137.

Larsen, K., M. B. Thygesen, F. Guillaumie, W. G. T. Willats, K. J. Jensen. 2006. Solid-phase chemical tools for glycobiology. *Carbohydrate Research* 341: 1209–34.

Lauber, M. A., Y. Q. Yu, D. W. Brousmiche, et al. 2015. Rapid preparation of released N-glycans for HILIC analysis using a labeling reagent that facilitates sensitive fluorescence and ESI-MS detection. *Analytical Chemistry* 87: 5401–9.

Li, L. L., J. Jiao, Y. Cai, Y. Zhang, H. J. Lu. 2015. Fluorinated carbon tag derivatization combined with fluorous solid-phase extraction: a new method for the highly sensitive and selective mass spectrometric analysis of glycans. *Analytical Chemistry* 87: 5125–31.

Liang, J., J. H. Zhu, M. M. Wang, et al. 2019. Evaluation of AGP fucosylation as a marker for hepatocellular carcinoma of three different etiologies. *Scientific Reports* 9: 11580.

Liu, X., H. Y. Qiu, R. K. Lee, W. X. Chen, J. J. Li. 2010. Methylamidation for sialoglycomics by MALDI-MS: A facile derivatization strategy for both alpha 2,3-and alpha 2,6-linked sialic acids. *Analytical Chemistry* 82: 8300–6.

Maley, F., R. B. Trimble, A. L. Tarentino, T. H. Plummer. 1989. Characterization of glycoproteins and their associated oligosaccharides through the use of endoglycosidases. *Analytical Biochemistry* 180: 195–204.

Martin, S. M., C. Delporte, A. Farrell, N. N. Iglesias, N. McLoughlin, J. Bones. 2015. Comparative analysis of monoclonal antibody N-glycosylation using stable isotope labelling and UPLC-fluorescence-MS. *Analyst* 140: 1442–7.

Mechref, Y., M. V. Novotny, C. Krishnan. 2003. Structural characterization of oligosaccharides using MALDI-TOF/TOF tandem mass spectrometry. *Analytical Chemistry* 75: 4895–903.

Mehta, N., M. Porterfield, W. B. Struwe, et al. 2016. Mass spectrometric quantification of N-linked glycans by reference to exogenous standards. *Journal of Proteome Research* 15: 2969–80.

Morrison, K. A., B. H. Clowers. 2017. Differential fragmentation of mobility-selected glycans via ultraviolet photodissociation and ion mobility-mass spectrometry. *Journal of the American Society for Mass Spectrometry* 28: 1236–41.

Mucha, E., M. Lettow, M. Marianski, et al. 2018. Fucose migration in intact protonated glycan ions: a universal phenomenon in mass spectrometry. *Angewandte Chemie-International Edition* 57: 7440–3.

Nishikaze, T. 2017a. Sensitive and structure-informative N-glycosylation analysis by MALDI-MS; Ionization, fragmentation, and derivatization. *Mass Spectrometry (Tokyo)* 6: A0060. doi: 10.5702/massspectrometry.A0060.

Nishikaze, T., H. Tsurnoto, S. Sekiya, S. Iwamoto, Y. Miura, K. Tanaka. 2017b. Differentiation of sialyl linkage isomers by one-pot sialic acid derivatization for mass spectrometry-based glycan profiling. *Analytical Chemistry* 89: 2353–60.

Nishikaze, T.. 2019. Sialic acid derivatization for glycan analysis by mass spectrometry. *Proceedings of the Japan Academy Series B-Physical and Biological Sciences* 95: 523–37.

North, S. J., H. H. Huang, S. Sundaram, et al. 2010. Glycomics profiling of Chinese hamster ovary cell glycosylation mutants reveals N-glycans of a novel size and complexity. *Journal of Biological Chemistry* 285: 5759–75.

Orlando, R., J. M. Lim, J. A. Atwood, et al. 2009. IDAWG: Metabolic incorporation of stable isotope labels for quantitative glycomics of cultured cells. *Journal of Proteome Research* 8: 3816–23.

Pabst, M., J. S. Bondili, J. Stadlmann, L. Mach, F. Altmann. 2007. Mass plus retention time = structure: A strategy for the analysis of N-glycans by carbon LC-ESI-MS and its application to fibrin N-glycans. *Analytical Chemistry* 79: 5051–7.

Pabst, M., J. Grass, S. Toegel, E. Liebminger, R. Strasser, F. Altmann. 2012. Isomeric analysis of oligomannosidic N-glycans and their dolichol-linked precursors. *Glycobiology* 22: 389–99.

Pang, P. C., P. C. N. Chiu, C. L. Lee, et al. 2011. Human sperm binding is mediated by the sialyl-lewis(x) oligosaccharide on the Zona pellucida. *Science* 333: 1761–4.

Peng, Y., J. Lv, L. J. Yang, D. Q. Wang, Y. Zhang, H. J. Lu. 2019a. A streamlined strategy for rapid and selective analysis of serum N-glycome. *Analytica Chimica Acta* 1050: 80–7.

Peng, Y., L. M. Wang, Y. Zhang, H. M. Bao, H. J. Lu. 2019b. Stable isotope sequential derivatization for linkage-specific analysis of sialylated N-glycan isomers by MS. *Analytical Chemistry* 91: 15993–6001.

Powell, A. K., D. J. Harvey. 1996. Stabilization of sialic acids in N-linked oligosaccharides and gangliosides for analysis by positive ion matrix-assisted laser desorption ionization mass spectrometry. *Rapid Communications in Mass Spectrometry* 10: 1027–32.

Qin, W. J., Z. J. Zhang, R. H. Qin, et al. 2019. Providing bionic glycome as internal standards by glycan reducing and isotope labeling for reliable and simple quantitation of N-glycome based on MALDI- MS. *Analytica Chimica Acta* 1081: 112–9.

Reiding, K. R., D. Blank, D. M. Kuijper, A. M. Deelder, M. Wuhrer. 2014. High-throughput profiling of protein N-glycosylation by MALDI-TOF-MS employing linkage-specific sialic acid esterification. *Analytical Chemistry* 86: 5784–93.

Reiding, K. R., A. Bondt, R. Hennig, et al. 2019. High-throughput serum N-glycomics: Method comparison and application to study rheumatoid arthritis and pregnancy-associated changes. *Molecular & Cellular Proteomics* 18: 3–15.

Ren, X. J., H. H. Bai, Y. T. Pan, et al. 2014. A graphene oxide-based immobilized PNGase F reagent for highly efficient N-glycan release and MALDI-TOF MS profiling. *Analytical Methods* 6: 2518–25.

Rohmer, M., B. Meyer, M. Mank, B. Stahl, U. Bahr, M. Karas. 2010. 3-amino-quinoline acting as matrix and derivatizing agent for MALDI MS analysis of oligosaccharides. *Analytical Chemistry* 82: 3719–26.

Royle, L., M. P. Campbell, C. M. Radcliffe, et al. 2008. HPLC-based analysis of serum N-glycans on a 96-well plate platform with dedicated database software. *Analytical Biochemistry* 376: 1–12.

Ruhaak, L. R., S. Miyamoto, K. Kelly, C. B. Lebrilla. 2012. N-glycan profiling of dried blood spots. *Analytical Chemistry* 84: 396–402.

Sandoval, W. N., F. Arellano, D. Arnott, H. Raab, R. Vandlen, J. R. Lill. 2007. Rapid removal of N-linked oligosaccharides using microwave assisted enzyme catalyzed deglycosylation. *International Journal of Mass Spectrometry* 259: 117–23.

Schwartz, E., Hayer, M., Hungate, B. A., Koch, B. J., McHugh, T. A., Mercurio, W., Morrissey, E. M., Soldanova, K. 2016. Stable isotope probing with O-18-water to investigate microbial growth and death in environmental samples. *Current Opinion in Biotechnology* 41: 14–18.

Sekiya, S., Y. Wada, K. Tanaka. 2005. Derivatization for stabilizing sialic acids in MALDI-MS. *Analytical Chemistry* 77: 4962–8.

Selman, M. H. J., M. Hemayatkar, A. M. Deelder, M. Wuhrer. 2011. Cotton HILIC SPE microtips for microscale purification and enrichment of glycans and glycopeptides. *Analytical Chemistry* 83: 2492–9.

Seo, Y., M. J. Oh, J. Y. Park, J. K. Ko, J. Y. Kim, H. J. An. 2019. Comprehensive characterization of biotherapeutics by selective capturing of highly acidic glycans using stepwise PGC-SPE and LC/MS/MS. *Analytical Chemistry* 91: 6064–71.

Shah, P., S. Yang, S. S. Sun, P. Aiyetan, K. J. Yarema, H. Zhang. 2013. Mass spectrometric analysis of sialylated glycans with use of solid-phase labeling of sialic acids. *Analytical Chemistry* 85: 3606–13.

Shajahan, A., N. Supekar, C. Heiss, P. Azadi. 2019. High-throughput automated micro-permethylation for glycan structure analysis. *Analytical Chemistry* 91: 1237–40.

Shubhakar, A., R. P. Kozak, K. R. Reiding, et al. 2016. Automated high-throughput permethylation for glycosylation analysis of biologics using MALDI-TOF-MS. *Analytical Chemistry* 88: 8562–9.

Snovida, S. I., V. C. Chen, H. Perreault. 2006. Use of a 2,5-dihydroxybenzoic acid/aniline MALDI matrix for improved detection and on-target derivatization of glycans: A preliminary report. *Analytical Chemistry* 78: 8561–8.

Song, T., D. Aldredge, C. B. Lebrilla. 2015. A method for in-depth structural annotation of human serum glycans that yields biological variations. *Analytical Chemistry* 87: 7754–62.

Song, X. Z., H. Ju, Y. Lasanajak, M. R. Kudelka, D. F. Smith, R. D. Cummings. 2016. Oxidative release of natural glycans for functional glycomics. *Nature Methods* 13: 528–34.

Song, W. R., X. M. Zhou, J. D. Benktander, et al. 2019. In-depth compositional and structural characterization of N-glycans derived from human urinary exosomes. *Analytical Chemistry* 91: 13528–37.

Stadlmann, J., M. Pabst, D. Kolarich, R. Kunert, F. Altmann. 2008. Analysis of immunoglobulin glycosylation by LC-ESI-MS of glycopeptides and oligosaccharides. *Proteomics* 8: 2858–71.

Staudacher, E., F. Altmann, I. B. H. Wilson, L. Marz. 1999. Fucose in N-glycans: From plant to man. *Biochimica Et Biophysica Acta-General Subjects* 1473: 216–36.

Stine, K. J. 2017. Application of porous materials to carbohydrate chemistry and glycoscience. *Advances in Carbohydrate Chemistry and Biochemistry* 74: 61–136.

Sun, D. H., F. L. Hu, H. Y. Gao, et al. 2019. Distribution of abnormal IgG glycosylation patterns from rheumatoid arthritis and osteoarthritis patients by MALDI-TOF-MSn. *Analyst* 144: 2042–51.

Suzuki, Y. 2005. Sialobiology of influenza molecular mechanism of host range variation of influenza viruses. *Biological & Pharmaceutical Bulletin* 28: 399–408.

Suzuki, Y. 2006. Highly pathogenic avian influenza viruses and their sialo-sugar receptors. *Trends in Glycoscience and Glycotechnology* 18: 153–5.

Szabo, Z., A. Guttman, B. L. Karger. 2010. Rapid release of N-linked glycans from glycoproteins by pressure-cycling technology. *Analytical Chemistry* 82: 2588–93.

Takasaki, S., T. Mizuochi, A. Kobata. 1982. Hydrazinolysis of asparagine-linked sugar chains to produce free oligosaccharides. *Methods in Enzymology* 83: 263–8.

Tao, S. J., Y. N. Huang, B. E. Boyes, R. Orlando. 2014. Liquid chromatography-selected reaction monitoring (LC-SRM) approach for the separation and quantitation of sialylated N-glycans linkage isomers. *Analytical Chemistry* 86: 10584–90.

Tousi, F., J. Bones, W. S. Hancock, M. Hincapie. 2013. Differential chemical derivatization integrated with chromatographic separation for analysis of isomeric sialylated N-glycans: A nano-hydrophilic interaction liquid chromatography-MS platform. *Analytical Chemistry* 85: 8421–8.

Toyoda, M., H. Ito, Y. K. Matsuno, H. Narimatsu, A. Kameyama. 2008. Quantitative derivatization of sialic acids for the detection of sialoglycans by MALDI MS. *Analytical Chemistry* 80: 5211–8.

Veillon, L., Y. F. Huang, W. J. Peng, X. Dong, B. G. Cho, Y. Mechref. 2017a. Characterization of isomeric glycan structures by LC-MS/MS. *Electrophoresis* 38: 2100–14.

Veillon, L., S. Zhou, Y. Mechref. 2017b. Quantitative glycomics: A combined analytical and bioinformatics approach. *Proteomics in Biology, Pt A* 585: 431–77.

Vreeker, G. C. M., S. Nicolardi, M. R. Bladergroen, et al. 2018. Automated plasma glycomics with linkage-specific sialic acid esterification and ultrahigh resolution MS. *Analytical Chemistry* 90: 11955–61.

Wang, Y. Y., Y. Cai, Y. Zhang, H. J. Lu. 2019. Glycan reductive amino acid coded affinity tagging (GRACAT) for highly specific analysis of N-glycome by mass spectrometry. *Analytica Chimica Acta* 1089: 90–9.

Wei, J., Y. Tang, M. E. Ridgeway, M. A. Park, C. E. Costello, C. Lin. 2020. Accurate identification of isomeric glycans by trapped ion mobility spectrometry-electronic excitation dissociation tandem mass spectrometry. *Analytical Chemistry* 92: 13211–20.

Wheeler, S. F., P. Domann, D. J. Harvey. 2009. Derivatization of sialic acids for stabilization in matrix-assisted laser desorption/ionization mass spectrometry and concomitant differentiation of alpha(2 -> 3)- and alpha(2 -> 6)-isomers. *Rapid Communications in Mass Spectrometry* 23: 303–12.

Yamagaki, T. 1999. A new analytical method for oligosaccharides and glyco-conjugates using post-source decay fragmentation by matrix-assisted laser desorption/ionization time-of-flight mass spectrometry. *Trends in Glycoscience and Glycotechnology* 11: 227–32.

Yan, S., J. Vanbeselaere, C. S. Jin, et al. 2018. Core richness of N-glycans of *Caenorhabditis elegans*: A case study on chemical and enzymatic release. *Analytical Chemistry* 90: 928–35.

Yang, S., H. Zhang. 2012. Solid-phase glycan isolation for glycomics analysis. *Proteomics Clinical Applications* 6: 596–608.

Yang, S., Yuan, W., Yang, W. M., Zhou, J.Y., Harlan, R., Edwards, J., Li, S. W., Zhang, H.2013. Glycan analysis by isobaric aldehyde reactive tags and mass spectrometry. *Analytical Chemistry* 85: 8188–95.

Yang, L.J., Du, X. X., Peng, Y., Cai, Y., Wei, L., Zhang, Y., Lu, H. J. 2019. Integrated pipeline of isotopic labeling and selective enriching for quantitative analysis of N-glycome by mass spectrometry. *Analytical Chemistry* 91: 1486–93.

Yuan, J. B., C. J. Wang, Y. J. Sun, L. J. Huang, Z. F. Wang. 2014. Nonreductive chemical release of intact N-glycans for subsequent labeling and analysis by mass spectrometry. *Analytical Biochemistry* 462: 1–9.

Zatsepin, T. S., D. A. Stetsenko, A. A. Arzumanov, E. A. Romanova, M. J. Gait, T. S. Oretskaya. 2002. Synthesis of peptide-oligonucleotide conjugates with single and multiple peptides attached to 2-aldehydes through thiazolidine, oxime, and hydrazine linkages. *Bioconjugate Chemistry* 13: 822–30.

Zauner, G., A. M. Deelder, M. Wuhrer. 2011. Recent advances in hydrophilic interaction liquid chromatography (HILIC) for structural glycomics. *Electrophoresis* 32: 3456–66.

Zhang, W., H. Wang, H. L. Tang, P. Y. Yang. 2011. Endoglycosidase-mediated incorporation of O-18 into glycans for relative glycan quantitation. *Analytical Chemistry* 83: 4975–81.

Zhang, W., W. Q. Cao, J. M. Huang, et al. 2015. PNGase F-mediated incorporation of O-18 into glycans for relative glycan quantitation. *Analyst* 140: 1082–9.

Zhang, Y., Y. Peng, Z. C. Bin, H. J. Wang, H. J. Lu. 2016. Highly specific purification of N-glycans using phosphate-based derivatization as an affinity tag in combination with Ti4+-SPE enrichment for mass spectrometric analysis. *Analytica Chimica Acta* 934: 145–51.

Zhang, Y., Y. Peng, L. J. Yang, H. J. Lu. 2018. Advances in sample preparation strategies for MS-based qualitative and quantitative N-glycomics. *Trac-Trends in Analytical Chemistry* 99: 34–46.

Zhang, Z. J., M. Westhrin, A. Bondt, M. Wuhrer, T. Standal, S. Holst. 2019. Serum protein N-glycosylation changes in multiple myeloma. *Biochimica Et Biophysica Acta-General Subjects* 1863: 960–70.

Zhang, L., C. Wang, Y. K. Wu, et al. 2020a. Microwave irradiation-assisted high-efficiency N-glycan release using oriented immobilization of PNGase F on magnetic particles. *Journal of Chromatography A* 1619: 460934.

Zhang, T., K. Madunic, S. Holst, et al. 2020b. Development of a 96-well plate sample preparation method for integrated N- and O-glycomics using porous graphitized carbon liquid chromatography-mass spectrometry. *Molecular Omics* 16: 355–63.

Zhao, X. Y., C. Guo, Y. Huang, et al. 2019. Combination strategy of reactive and catalytic matrices for qualitative and quantitative profiling of N-glycans in MALDI-MS. *Analytical Chemistry* 91: 9251–8.

Zhou, H., A. C. Briscoe, J. W. Froehlich, R. S. Lee. 2012. PNGase F catalyzes de-N-glycosylation in a domestic microwave. *Analytical Biochemistry* 427: 33–5.

Zhou, S. Y., Y. L. Hu, L. Veillon, et al. 2016a. Quantitative LC-MS/MS glycomic analysis of biological samples using aminoxyTMT. *Analytical Chemistry* 88: 7515–22.

Zhou, S. Y., N. Tello, A. Harvey, B. Boyes, R. Orlando, Y. Mechref. 2016b. Reliable LC-MS quantitative glycomics using iGlycoMab stable isotope labeled glycans as internal standards. *Electrophoresis* 37: 1489–97.

Zhou, S. Y., X. Dong, L. Veillon, Y. F. Huang, Y. Mechref. 2017a. LC-MS/MS analysis of permethylated N-glycans facilitating isomeric characterization *Analytical and Bioanalytical Chemistry* 409: 453–66.

Zhou, S. Y., Y. F. Huang, X. Dong, et al. 2017b. Isomeric separation of permethylated glycans by porous graphitic carbon (PGC)-LC-MS/MS at high temperatures. *Analytical Chemistry* 89: 6590–7.

Zhou, S. Y., L. Veillon, X. Dong, Y. F. Huang, Y. Mechref. 2017c. Direct comparison of derivatization strategies for LC-MS/MS analysis of N-glycans. *Analyst* 142: 4446–55.

CHAPTER 4

Qualitative and Quantitative Analytical Methods for Intact Glycopeptides

Weiqian Cao and Pengyuan Yang

CONTENTS

DOI: 10.1201/9781003185833-4

4.1 OVERVIEW OF GLOBAL ANALYSIS OF INTACT GLYCOPEPTIDES

Great strides have been made in MS-based glycomic and deglycosylation-centric glycoproteomic studies in recent decades, which have greatly expanded our knowledge of protein glycosylation (Varki 2017). Further questions about the glycan compositions/structures at specific glycosylation sites have been raised. Consequently, MS-based intact glycopeptide analysis (also known as site-specific glycan analysis) has emerged as a more challenging issue in glycoproteomics, with the goal of identifying glycopeptide sequences with an attached glycan. Despite the great difficulties in intact glycopeptide analysis, various efforts towards elucidation of site-specific glycosylation on proteins have been made. Many analytical strategies, such as glycopeptide purification and separation, MS acquisition, and mass spectra interpretation, have been indicated to be effective and are generically available for extensive characterization of intact glycopeptides.

4.1.1 Comprehending the Glycosylation Process and Its Characteristics Is Essential for Intact Glycopeptide Analysis

The diversity and complexity of site-specific glycosylation bring difficulties for intact glycopeptide analysis (Hart and Copeland 2010). Understanding the protein glycosylation process and its characteristics is favourable for developing methods for intact glycopeptide analysis. The complicated structural properties of glycans attached to proteins arise from a series of competing enzymatic steps through non-template-driven intricate biosynthetic processes. It has been estimated that there are more than 7000 structures of mammalian glycans that are assembled from only 10 monosaccharides: mannose (Man), fucose (Fuc), sialic acid (SA), xylose (Xyl), glucose (Glc), galactose (Gal), N-acetylgalactosamine (GalNAc), N-acetylglucosamine (GlcNAc), glucuronic acid (GlcA), and iduronic acid (IdoA). Approximately, 700 proteins, including ~200 glycosyltransferases, participate in protein glycosylation pathways and generate diverse mammalian glycans (Moremen et al. 2012). Glycans are normally divided into different subtypes according to the method by which they are attached to polypeptides. N-glycans and O-glycans are the main glycan subtypes and have been investigated for their important biological roles.

N-glycans are first synthesized in the endoplasmic reticulum on a dolichol, ultimately creating a high-mannose structure with nine mannoses and a triglucose terminus. The high-mannose structure is then transferred to a nascent polypeptide and helps guide the protein folding process. The high-mannose structure attached to a protein further undergoes disassembly and rebuilding processes regulated by a series of glycotransferases in the Golgi to yield a variety of N-glycan structures. Typically, N-glycans contain a core structure of GlcNAc2Man3 and can be classified into three groups based on the different glycan composition and branching patterns: high-mannose type, complex type, and hybrid type (Figure 4.1A). N-glycans are attached to the protein through amide linkages to asparagine (N) side chains and normally with a consensus

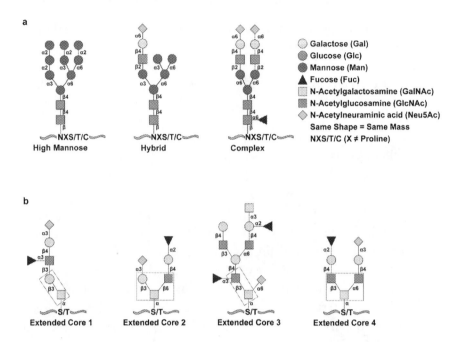

FIGURE 4.1 The complexity and variation of N- and O-glycopeptides. (A) Three types of N-glycans. (B) Four core structures of mucin-type O-glycans.

sequence of N-X-S/T, where X is any amino acid except proline. Other motifs, such as N-X-C/V/A (X≠P) and N-G, are also identified in organisms (Zielinska et al. 2010).

O-glycosylation modification is also performed in the endoplasmic reticulum and the Golgi. However, in contrast to N-glycosylation, O-glycosylation features a carbohydrate residue covalently linked to the hydroxyl group of the amino acids serine, threonine, and tyrosine. The monosaccharides of Fuc, Glc, GalNAc, GlcNAc, Man, and Xyl can be directly linked to the amino acid without a unique consensus motif such as N-glycosylation. O-glycosylation includes several types: O-fucosylation, O-glucosylation, O-GalNAcylation or mucin-type glycosylation, O-GlcNActylation, O-mannosylation, and O-xylosylation (Darula and Medzihradszky 2018). In particular, O-GalNAcylation is the most common mammalian O-glycosylation. The O-GalNAc glycans of mucins have four major core structures (Figure 4.1B). In addition, mucin-type glycosylation typically occurs on a consecutive serine or threonine, which greatly diversifies the O-glycopeptide composition. The diverse glycan structures resulting from the complexity of the glycosylation process endow glycoproteins with considerable variability and bring difficulties in glycoproteomic analysis (You et al. 2018).

4.1.2 The Status and Main Challenges of Intact Glycopeptide Analysis

Despite variances among the glycosylation processes, the resulting site-specific glycans represent very similar analytical challenges (Cao et al. 2020). First, the abundance of glycopeptides is inherently low in biological samples, and the diversity of glycosylation makes one glycopeptide much lower in stoichiometry. Second, the dissociation behaviours of glycans and peptides in mass spectrometry (MS) analysis are different. It is difficult to acquire informative glycopeptide spectra using a single MS fragmentation method. Additionally, the dissociation efficiency of glycopeptides is very low compared to that of nonglycosylation peptides, which could result in the signal suppression of glycopeptides in the MS analysis by nonglycosylation peptides. Third, a protein could have multiple glycosylation sites, and different glycoforms could occur on the same site, which greatly diversifies the glycopeptide composition. The glycopeptides comprising various types of monosaccharides linked to amino acid residues with different compositions and linkages are too complicated to directly analyze by routine proteomic strategies.

In recent years, especially in the past two decades, substantial progress has been achieved in addressing each of the above barriers. MS-based methods have been developed for global analysis of intact glycopeptides, such as glycopeptide enrichment and separation methods for improving the abundance of glycopeptides and reducing the signal suppression of glycopeptides by nonglycopeptides in MS analysis, mass spectral acquisition methods for obtaining informative glycopeptide spectra data, and software tools for spectral interpretation.

In a typical bottom-up intact glycopeptide identification strategy (Figure 4.2), a routine proteome pipeline was employed to obtain the digested peptides. For N-glycopeptide analysis, glycopeptides are enriched from the digested peptides with or without pre-fractionation and directly analyzed by MS. For O-glycopeptide analysis, peptides were treated by Peptide-N-glycosidase F (PNGase F) or other methods to remove the N-glycans, and then enriched and analyzed by MS. Different enrichment methods and analyzing strategies have been developed.

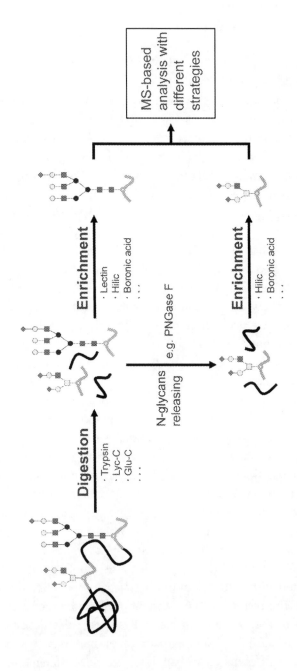

FIGURE 4.2 A typical bottom-up workflow for intact glycopeptide identification.

4.2 PRETREATMENT TECHNIQUES IN MS-BASED INTACT GLYCOPEPTIDE ANALYSIS

4.2.1 Sample Preparation

Similar to shotgun proteomics, the first step of sample preparation for intact glycopeptide analysis is protein extraction from biological samples, followed by digestion of proteins into peptides using proteolytic enzymes, such as trypsin, lysine C or multiple different enzymes. Efficient protein extraction and specific digestion of proteins are fundamental for intact glycopeptide analysis (Thaysen-Andersen and Packer 2014). Denaturing agents that are normally used for protein extraction in proteomic analysis, such as sodium dodecyl sulfate (SDS), urea, and NP40, are also used in intact glycopeptide analysis. SDS is preferred, given its benefits in cell membrane protein extraction where abundant glycoproteins are located. The selection of denaturing agents should take the subsequent processing into account. For example, NP40 is usually selected given its compatibility with SDS-gel separation or immunoprecipitation. Serum is usually used as a mild denaturing method (diluted in 50 mM ammonium bicarbonate solution, pH 8.5 and denatured by heating at 100°C for 30 seconds and cooling on ice for 30 seconds alternatively) to avoid protein flocculation and precipitation. Trypsin is also the most common enzyme for protein digestion in intact glycopeptide analysis. Studies have performed optimization experiments during sample preparation, including denaturing agent selection, protein storage time, trypsin type, enzyme-to-substrate ratio, and protein concentration, to achieve better intact glycopeptide identification results. Other enzymes, such as Lys-C and Glu-C, were also used to improve sequence coverage in MS glycopeptide identification. PNGase F, which is commonly used for N-glycan release, is typically used for the removal of N-glycans prior to glycopeptide enrichment to avoid the signal suppression from N-glycopeptides in MS O-glycopeptide analysis.

4.2.2 Enrichment Methods for Intact Glycopeptides

Proteins digested with trypsin or other proteolytic enzymes produce multiple peptides. The resulting peptide mixtures can be very complex and the abundance of glycopeptides is extremely low (approximately 1% of the peptide mixtures). Thus, prior to MS analysis, a glycopeptide enrichment procedure is almost always required, which can minimize the interference and signal suppression from highly abundant nonglycopeptides in

MS analysis. Different from the enrichment methods for deglycosylation-centric glycopeptide analysis, a necessary enrichment method is retrieving intact glycopeptides (Xiao et al. 2019). To this end, a variety of methods have been developed to enrich glycopeptides, each of which has its own advantages and limitations (Figure 4.3, Table 4.1).

4.2.2.1 Lectin Affinity-Based Enrichment

Lectins are a class of proteins, each of which can recognize specific types of carbohydrates. Lectins are responsible for cell surface sugar recognition in bacteria, animals, and plants. Examples include bacterial toxins, animal receptors, and plant toxins and mitogens. There are two key features in lectin-carbohydrate recognition: relatively high selectivity and relatively low affinity.

Glycopeptide selectivity is achieved through a combination of hydrogen bonding to the sugar hydroxyl groups with van der Waals packing and hydrophobic interactions. Each lectin is specific for only one or two types of monosaccharides. For example, concanavalin A (Con A) can predominately recognize alpha-mannose, which is very common in N-linked glycans. Wheat germ agglutinin (WGA) can recognize GlcNAc and N-acetylneuraminic acid (sialic acid). Aleuria aurantia lectin (AAL) can recognize the complex-type oligosaccharides with an a-fucosyl residue at the innermost GlcNAc residue (Osawa and Tsuji 1987). Lectins are usually used alone or in combination to enrich glycopeptides with unique glycan motifs. After lectin-glycopeptide binding, the bound fraction containing the enriched glycopeptides is subsequently eluted by employing specific mono- or disaccharides that compete with the glycopeptide, or by employing acidic conditions that disrupt the glycopeptide-lectin interaction. The use of lectins makes up for the insufficiency of MS, which has difficulty in discriminating structural isomers. For example, galectins, a group of galactose-binding proteins, can discriminate mannose from galactose. Siglecs, a group of sialic acid-binding lectins, can discriminate various linkage isomers of sialylated glycans. In recent decades, this carbohydrate-binding capacity has been widely exploited for the isolation of glycopeptides from complex samples (Dai et al. 2009). Lectins can also be used to fractionate both N- and O-linked oligosaccharides of glycoproteins through specific types of monosaccharide binding (Durham and Regnier 2006).

To simplify the enrichment operation, lectins are typically immobilized onto a solid support serving for solid-phase extraction of glycopeptides

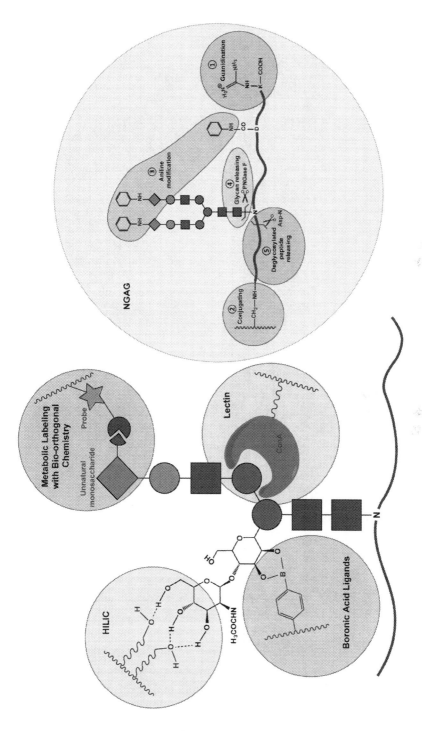

FIGURE 4.3   Enrichment strategies for intact N-glycopeptides.

TABLE 4.1   Characteristics of Different Enrichment Strategies for Intact Glycopeptides

Enrichment Methods	Advantages	Limitations
Lectin-affinity-based methods	Relative high selectivity; excellent ability to recognize different types of monosaccharides	Low specificity; biased toward glycopeptides
Boronic-acid-based methods	Reversible covalent reaction through pH controlling; unbiased enrichment method	Easily interfered by those compounds which contain 1–2 and 1–3 cis-diol groups
HILIC	Unbiased toward varied glycopeptides; high throughput	Relative low specificity
ZIC-HILIC	High selectivity; high specificity; high throughput	Relative low efficiency for glycopeptides with short glycans
Isotope-targeted glycoproteomics	Smart design; high specificity	Complex operation step; uncertain metabolizing efficiency
NGAG	Comprehensive characterization of glycoproteins	Tedious operation procedure; uncertain reaction efficiency

(Monzo et al. 2007). The lectin-functionalized beads can also be packed into a column (Kaji et al. 2003). However, the affinity of the lectins for monosaccharides is generally weak ($K_d = 10^{-4}$–10^{-7} M) compared with antigen-antibody interactions ($K_d = 10^{-8}$–10^{-12} M). Relatively low affinity is caused by shallow indentations on protein surfaces (Weis and Drickamer 1996).

4.2.2.2 Boronic Acid-Based Enrichment

Boronic acid-based enrichment is based on the reversible covalent reaction between boronic acid ligands and cis-diol-containing compounds (Li et al. 2015). The saccharides linked to glycopeptides contain cis-diol groups. During the enrichment procedure, cyclic boronate esters are selectively formed at high pH and the reaction can be reversed under acidic conditions.

Thus, boronic acid chemistry can be employed for the enrichment of glycopeptides containing saccharides such as mannose, galactose, or glucose. In contrast to lectins, the interaction between the glycopeptide and boronic acid does not require a complex recognition motif consisting of several saccharides. This feature has made boronic acid chemistry an unbiased enrichment method of both N- and O-glycopeptides. Since this method is based on covalent interactions, the bonding strength is stronger than that of hydrogen interactions or van der Waals interactions. Furthermore, the capture/release can be easily controlled through a simple switch of the pH and the acidic solutions required for elution are compatible with MS.

A variety of new boronate affinity materials, such as monoliths, mesoporous silica, magnetic particles, or gold nanoparticle-based materials have been functionalized with boronic acid, demonstrating high specificity for glycopeptides (Xiao et al. 2018). Moreover, numerous efforts have been made to solve fundamental issues, such as binding pH, affinity, and selectivity. On the other hand, the specificity for glycopeptide enrichment may be depressed by compounds that contain 1-2 and 1-3 cis-diol groups, such as nucleotides from complex mixtures. Since most boronic acid derivatives possess a pKa of 8–9 and the pH of frequently used biosamples ranges from 4.5 to 8.0, the use of a basic pH necessitates the inconvenience of pH adjustment and is associated with the risk of degradation of labile biomolecules.

4.2.2.3 HILIC-Based Enrichment

Hydrophilic-interaction liquid chromatography (HILIC), which is a variant of normal phase liquid chromatography, was introduced by Alpert A. J. in 1990 (Alpert 1990). HILIC is capable of separating polar and hydrophilic solutes with a polar stationary phase and a less polar mobile phase. In general, water-acetonitrile is typically the optimal choice for the mobile phase. HILIC-mediated separation of glycopeptides from mixtures mainly depends on the hydrophilic difference between glycopeptides and non-modified peptides, given the more hydrophilic characteristics of glycans.

HILIC has remarkable advantages of no biases towards varied glycopeptides, satisfactory compatibility with MS, high throughput and low cost. Among all the factors that affect the separation performance, the HILIC material directly determines the selectivity of solutes. In recent years, the glycopeptide enrichment selectivity of HILIC has been improved significantly (Qing et al. 2020).

Zwitterionic HILIC (ZIC-HILIC) is a type of HILIC commercially used in glycopeptide enrichment. Zwitterion has both cationic and anionic moieties in one molecule and can maintain overall charge neutrality. Zwitterion possesses a strong hydration capacity. Given their excellent hydrophilicity and biocompatibility, ZIC-based HILIC columns can efficiently separate hydrophilic and polar substances in a hydrophilic mode and are widely used for glycopeptide enrichment. Moreover, glycopeptides can be better analyzed by utilizing ZIC-HILIC and monolithic RP columns sequentially, which take advantage of both high selectivity and excellent separation ability (Jiang et al. 2019). In addition, carbohydrate-functionalized HILIC materials, amide- or amine-functionalized HILIC materials and other HILIC materials have been developed (Xiao et al. 2018).

4.2.2.4 Other Methods

There are many other recently developed enrichment methods that integrate glycopeptide pretreatment or identification procedures.

Isotope-targeted glycoproteomics (IsoTaG) is a mass-independent platform for intact N- and O-glycopeptide discovery and analysis (Woo et al. 2015). Artificial synthetic glycans such as $Ac_4GalNAz$ and $Ac_4ManNAz$ are first metabolically labelled into cells to replace the corresponding GalNAc, GlucNAc, and SiaNAc on N- and O-glycoproteins. Then, these tryptic digested glycopeptides are enriched/labelled using an artificial isotopic recoding affinity probe by click reaction between probes and chemically functionalized glycans. The labelled glycopeptides are identified with direct tandem MS and targeted glycopeptide assignment. With this method, the glycan structure and peptide can be identified for both intact N- and O-glycopeptides.

Solid-phase extraction of N-linked glycans and glycosite-containing peptides (NGAG) is a chemoenzymatic method for the comprehensive characterization of glycoproteins that is able to determine glycan heterogeneity for individual glycosites in addition to providing information about the total N-linked glycan, glycosite-containing peptide, and glycoprotein content of complex samples (Sun et al. 2016).

Protein samples were first digested into peptides followed by guanidination to block the ε-amino groups on the side chain of lysine residues. Then the peptides were covalently conjugated to an aldehyde-functionalized solid support through their N-termini (α-amino groups). Afterwards, the carboxyl groups of aspartic acid, glutamic acid, and peptide C-termini were modified with aniline. Sialic acids on the sialylated glycopeptides were also simultaneously modified by aniline to facilitate mass spectrometric detection of sialylated glycans. N-linked glycans were released from the solid support by PNGase F digestion in parallel with conversion of asparagine residues at N-linked glycosites to aspartic acid residues by deamination. Glycosite-containing peptides with aspartic residues at N-glycosites were specifically released from the solid phase by Asp-N digestion. Finally, the released N-glycans and glycosite-containing peptides were identified by MS, and the intact N-glycopeptides were reconstructed from the analysis results of N-glycans and glycosite-containing peptides.

Metabolic labelling is good at monitoring dynamically generated glyco-proteins (Prescher et al. 2004). Isotope-labelled saccharides are first meta-bolically incorporated into glycoproteins, and the resulting glycopeptides can be analyzed by LC-MS/MS (Xu et al. 2019). In recent years, metabolic labelling with bioorthogonal chemistry has been used in glycopeptide enrichment. Unnatural sugars containing bioorthogonal functional groups are metabolically labelled on glycoproteins and then enriched by bioor-thogonal chemistry. This method has high specificity and provides an effective method for the large-scale analysis of glycopeptides in vivo.

4.2.3 Separation Methods for Intact Glycopeptides

Separation of glycopeptides through liquid chromatography (LC) prior to MS analysis is necessary to reduce the complexity of glycopeptides and improve the MS identification efficiency. Different chromatographic separation techniques have been applied to glycopeptide analysis, such as reversed-phase liquid chromatography (RPLC), HILIC, and strong cation exchange chromatography (SCX). The LC methods are usually coupled with MS analysis as in proteomic studies or used independently to reduce the complexity of samples. RPLC is normally coupled with an MS instru-ment for the LC-MS/MS analysis of glycopeptides. RPLC uses hydropho-bic materials such as C8 or C18-bonded silica as the stationary phase to retain peptides. Although the strength interaction between an intact gly-copeptide and the hydrophobic stationary phase is mainly determined by the properties of peptides, microheterogeneity exists in retention times of glycopeptide. Other chromatography methods, such as HILIC and SCX, are commonly used independently prior or after glycopeptide enrichment as two-dimensional (2D) LC-MS/MS analysis, which can greatly increase the identification results (Jiang et al. 2019).

4.3 MS METHODS FOR GLOBAL ANALYSIS OF INTACT GLYCOPEPTIDES

4.3.1 Mass Spectrometry Instrumentation

Intact glycopeptide analysis generally requires high resolution for MS instruments. High resolution enables more specific identification and elucidation of glycoforms. MALDI-TOF, Q-TOF, Fourier-transform ion cyclotron resonance (FTICR), and orbitrap MS have all been used for intact glycopeptide analysis. Intact glycopeptides from standard glycoproteins, which do not need extensive separation, may be obtained by direct MALDI-TOF analysis based on the known molecular weight of these glycopeptides. Although FTICR has been used for glycopeptide analysis, its slow analysis is a limitation; thus, it is not suitable for high-throughput glycopeptide analysis. Orbitrap MS demonstrates the most effective and best performance in high-throughput glycopeptide analysis. Q-TOF has high scan rates and is essential in glycopeptide analysis. Different fragmentation methods on MS instruments have their own characteristics for intact glycopeptide analysis.

4.3.2 MS Fragmentation Methods for Intact Glycopeptides

An appropriate MS fragmentation method to obtain informative glycopeptide mass spectra is essential in glycopeptide analysis. Intact glycopeptide fragmentation is complicated and involves glycan fragmentation and peptide fragmentation. However, the dissociation behaviours of glycans and peptides in glycopeptide fragmentation are different because of the differences in the physical and chemical properties of the glycosidic bonds and peptide bonds. Glycans exhibit significantly lower intrinsic basicity and proton affinity than peptides. Glycosidic bonds are more liable than peptide bonds, and glycans tend to be more easily fragmented than peptide parts. Glycans primarily produce B and Y ions, while peptides generally produce b/y or z/c ions as a result of glycopeptide fragmentation (Figure 4.4).

Different fragmentation mechanisms have become available on commercial MS instruments and are used for specialized strategies for glycopeptide analysis due to their own characteristics (Figure 4.5, Table 4.2) (Hu et al. 2017; Thaysen-Andersen et al. 2016). These fragmentation mechanisms, including resonance activation (ion trap) collision-induced dissociation (CID), beam-type (Q-TOF) CID, higher-energy collision dissociation (HCD), and electron-transfer dissociation (ETD), are the most commonly used approaches in glycopeptide identification.

4.3.2.1 CID Fragmentation

Ion trap CID is a typical low-energy dissociation method in glycoproteomic studies. Collisions with neutral gas molecules and surfaces are two main types of CID methods but yield the same fragmentation results (Dongre and Wysocki 1994). CID with neutral gas molecules, which are

a

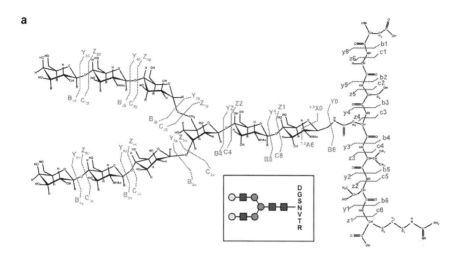

b

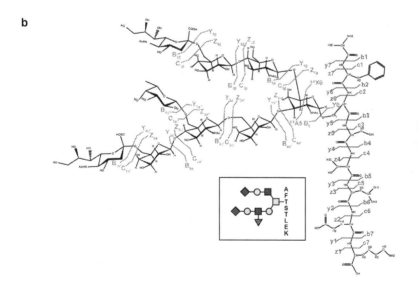

FIGURE 4.4 Examples of N-glycopeptide (A) and O-glycopeptide (B) fragmentation nomenclature.

typically employed in ion trap MS analyzers, is common in glycopeptide analysis. Since glycan fragmentations are low-energy processes, Y ions and B ions from glycans are more easily fragmented during the CID process. However, due to the "1/3" cutoff in the low m/z region (Hu et al. 2016), the B ions are mostly excluded in the mass spectra generated by ion trap CID.

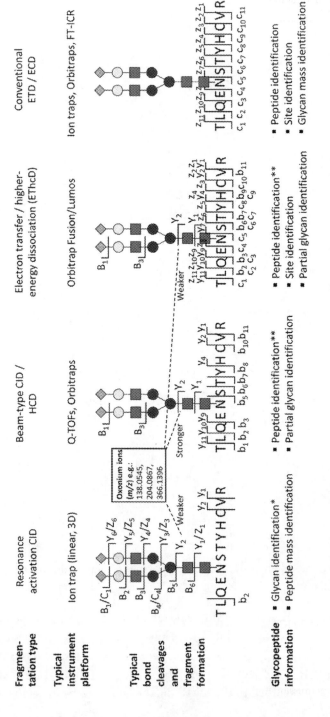

FIGURE 4.5 The main dissociation methods in glycoproteomics, typically used in conjunction i.e., resonance activation (ion trap type) CID, beam-type (Q-TOF type) CID or HCD, hybrid-type EThcD, and the conventional ETD (or ECD) fragmentation. *Reprinted with permission from ref.* Thaysen-Andersen, M. et al. 2016. *mol cell proteomics, 15(6): 1773–90.*

TABLE 4.2   *Structural Information Provided by Different Fragmentation Methods

Fragmentation Method	Instrument Type	Type of Fragment Ions Generated
Trap CID	IT	• Stub-glycopeptide ions (intact peptide ions with small glycan fragments) • Abundant ions from loss of monosaccharide units from the precursor • Oxonium ions are observed based on acquisition range and precursor m/z • Peptide backbone ions (b and y) may be observed in sequential tandem MS
Beam type CID	TOF-TOF, Q-TOF, Q-FTICR	• Mono-, di-, or tri-saccharide oxonium ions (glycosidic bond cleavages) • Stub-glycopeptide ions • Ions from loss of monosaccharide units from the precursor • Peptide backbone ions (b and y) • Peptide backbone (b and y) ions with the starting monosaccharide (HexNAc for N-linked glycopeptides)
HCD	Orbitrap	• Mono-, di-, or tri-saccharide oxonium ions (glycosidic bond cleavages) • Stub-glycopeptide ions • Ions from loss of monosaccharides from the precursor • Peptide backbone ions (b and y)
ECD	FTICR	• Peptide backbone ions (c and z ions with modification) • Labile PTMs like glycans and phosphate groups remain largely intact at the modification site • Charge remote fragmentations, such as internal fragments and amino acid side chain losses can occur • Better for shorter peptides than ETD due to additional fragmentation, which can be controlled by varying average electron energy
ETD	IT-FTICR, IT-Orbitrap	• Same as ECD (c and z ions with modifications intact) • Prone to steric effects and electron affinity depending on reagent used • Fewer charge remote fragmentations and secondary cleavages than in ECD
UVPD	IT-FTICR, IT-Orbitrap	• Peptide backbone ions with labile modification intact (predominantly a and x ions) • Diagnostic ions from glycosidic bond and cross-ring cleavages

*Reprinted with permission from reference Hu H. et al. (2017). *Mass Spectrometry Reviews* 36(4):475–98.

Peptides have difficulties in producing abundant fragments under CID conditions. Thus, Y ions as well as a few B ions and a limited number of b/y ions are dominant in the mass spectra generated by ion trap CID. Due to the characteristics of CID-generated mass spectra, the conjugated glycans could be confidently identified through the CID-induced method, whereas the location of the glycosylation site and the peptide sequences could not. Thus, CID fragmentation is rarely used alone for glycopeptide analysis, especially in complex samples and is only applied to analyze glycopeptides in simple samples in some previous studies.

4.3.2.2 HCD Fragmentation

The HCD on the orbitrap mass spectrometer fragmentation of glycopeptides is highly similar to that of beam-type CID on Q-TOF instruments. This kind of fragmentation overcomes the 1/3 m/z cutoff associated with ion trap CID. Glycan oxonium ions (the common glycan oxonium ions are summarized in Table 4.3), such as m/z 138.055, 204.087,

TABLE 4.3　*The Common Oxonium Ions from Glycopeptides.

Oxonium Ions (m/z)	Fragment Type	Chemical Formula
109.028	Hex fragment	$C_6H_4O_2$
115.039	Hex fragment	$C_5H_6O_3$
126.055	HexNAc fragment	$C_6H_7O_2N_1$
127.039	Hex-2H$_2$O	$C_6H_6O_3$
138.055	HexNAc fragment	$C_7H_7O_2N_1$
144.066	HexNAc fragment	$C_6H_9O_3N_1$
145.05	Hex-H$_2$O	$C_6H_8O_4$
163.06	Hex	$C_6H_{10}O_5$
168.066	HexNAc-2H$_2$O	$C_8H_9O_3N_1$
186.076	HexNAc-H$_2$O	$C_8H_{11}O_4N_1$
204.087	HexNAc	$C_8H_{13}O_5N_1$
274.092	NeuAc-H$_2$O	$C_{11}H_{15}O_7N_1$
290.087	NeuGc-H$_2$O	$C_{11}H_{15}O_8N_1$
292.103	NeuAc	$C_{11}H_{17}O_8N_1$
308.098	NeuGc	$C_{11}H_{17}O_9N_1$
366.14	Hex+HexNAc	$C_{14}H_{23}O_{10}N_1$
657.14	Hex+HexNAc+NeuAc	$C_{25}H_{40}O_{18}N_2$
673.23	Hex+HexNAcNeuGc	$C_{25}H_{40}O_{19}N_2$

*Translated with permission from reference Wen-Feng Zeng et al. (2016) *Progress in Biochemistry and Biophysics* 43 (6): 550–62.

and 366.140, are detectable and can be used as diagnostic ions to indicate the presence of glycopeptides (Zeng et al. 2016b). In addition, distinct Y1 ions (peptide+GlcNAc) that are very useful in glycopeptide assignment can be generated. However, glycan-type or peptide-type fragments tend to be produced in different collision energies in this fragmentation mode. At low energy, the method tends to yield B and Y ions for glycan identification, but produces more b/y-ion series for the identity of peptides at the expense of glycan characterization at a high-energy level. Stepped collision energies (SCE)-HCD employed in Thermo mass spectrometers can produce complementary fragments of glycans and peptides in a single spectrum, providing much more needed information on the identification of intact glycopeptides. SCE-HCD at 20–30–40% generates the most informative and abundant fragment ions of glycans and peptides and has been widely used as an optimized collision energy in HCD fragmentation for intact glycopeptide analysis (Liu et al. 2017b).

4.3.2.3 Electron-Based Fragmentation

Electron-capture dissociation (ECD) and ETD are two main electron-based fragmentation techniques in glycopeptide analysis (Zhu and Desaire 2015). ECD is performed traditionally in FTICR mass spectrometers, whereas ETD works on ion trap instruments. In ECD and ETD, a glycopeptide forms a radical precursor ion and undergoes fragmentation via cleavages of N-C_α bonds, primarily yielding c/z ions generated from the peptide backbone on the glycopeptide while the remaining glycan part is intact. The particular dissociation pathway of ECD/ETD makes these methods ideal for identification of peptide sequences and assignment of glycosylation sites. However, these methods often suffer from incomplete fragmentation, leading to a large amount of residual precursor ions. Studies have been performed to increase the efficiency of glycopeptide dissociation in ECD/ETD. For example, in ECD mode, increasing the electron energy from 0.2 eV to greater than 4 eV greatly improved the sequence coverage of the glycopeptide. It was found that the efficiency of glycopeptide dissociation in ETD increases when the glycopeptide ion carries more charges. Thus, charging reagents, such as m-nitrobenzyl alcohol, were employed as elution buffer in LC separation before MS analysis to increase the average charge state of glycopeptides.

4.3.2.4 Photon-Induced Dissociation

In photon-induced dissociation, the internal energy of a precursor is increased by absorbing photons to induce fragmentation (Brodbelt 2014). Infrared multiphoton dissociation (IRMPD) is a common dissociation method of this type that uses CO_2 lasers to irradiate infrared photons at micrometer wavelengths with energetics similar to CID, in which glycosidic bonds are preferentially cleaved, yielding fragments similar to CID. Ultraviolet photodissociation (UVPD), which relies on ion-photon interactions, is another promising tool for elucidating glycopeptide structures. In UVPD, the internal energy of the ion was excited to higher electronic states of approximately 3.5–7.9 eV by the UV photons generated from lasers. In this procedure, in addition to excitation, energy redistribution could occur before or during ion dissociation, resulting in a new fragmentation pathway and generating more diverse fragments. For example, both glycosidic and cross-ring cleavages occur in UVPD, resulting in more A- and X-type ions that are useful for elucidating glycan structures. Two wavelengths in UVPD, 157 and 193 nm are commonly used for analyzing glycopeptides.

4.3.2.5 The Combination of Fragmentation Method

The glycopeptide fragmentation behaviours have been extensively investigated under different dissociation methods (Aboufazeli et al. 2015; Hinneburg et al. 2016). Different dissociation methods have their own characteristics and strengths. Most of these methods are complementary for glycopeptide fragmentation in LC-MS/MS analysis. For instance, electron-based fragmentation is highly complementary to the dissociation pathway in CID/HCD. Hybrid dissociation techniques, such as ETciD and EThcD, in which a supplementary energy CID/HCD source is applied to all the ETD-generated ions, have shown great potential in glycopeptide analysis. Activated-ion electron transfer dissociation (AI-ETD), which uses concurrent infrared photoactivation to improve dissociation efficiencies and increase product ion generation in ETD reactions, is particularly well-suited for intact glycopeptide fragmentation and can enable large-scale glycoproteomic analysis (Riley et al. 2019). HCD-product-dependent (pd)-ETD is another type of fragmentation method that uses the product ion trigger technique. Generally, further data-dependent scans with ETD were triggered only with the detection of a specified product ion from a given mass list, which are normally glycan oxonium ions that are easily detectable and can be used as diagnostic ions to indicate the presence of glycopeptide. This kind of combination method largely avoids the reaction time of ETD wasted on nonglycopeptides dissociation.

4.3.3 Bottom-Up Experimental and MS Acquisition Strategies for Intact N-Glycopeptide Analysis

Bottom-up experimental strategies were designed for intact N-glycopeptide analysis with the principle of acquiring informative mass spectra for glycopeptide interpretation. For easy understanding, we divided the current strategies into two main types according to whether an extra identification of peptide segments from glycopeptides is conducted or not (Figure 4.6).

4.3.3.1 Experimental Strategy with Extra Peptide Identification
In the experimental strategy with extra peptide identification, the identification of peptide segments from glycopeptides is needed. Normally, an MS3 acquisition step or a deglycosylation procedure for pre-establishing a peptide database is introduced for the identification of the peptide segments. The main purpose of both the MS3 step and the deglycosylation procedure is to deduce the carrier peptides from the intact glycopeptides.

The strategy with an MS3 acquisition step normally assigns the conjugated glycan composition by CID-MS/MS that is adept at producing Y ions and defines the carrier peptide sequences by a further targeted MS3 analysis. So far, two strategies, namely Sweet Heart proposed by Wu et al. (Wu et al. 2013) and pGlyco proposed by Zeng et al. (Zeng et al. 2016a), followed this principle. A specially designed strategy, isotope-targeted glycoproteomics (IsoTag) (Woo et al. 2015), which also uses MS3 for peptide identification, follows a novel pipeline. This approach smartly uses the natural abundance of the stable isotopes ^{79}Br and ^{81}Br (1:1) to provide a ready source for isotope recoding and convoluting the mass envelope of a glycopeptide with a triple signature (1:2:1 [M, M+2, M+4]), which is used to trigger the MS2 analysis for glycoforms, applying MS3 analysis to identify the peptide isoforms.

Generally, in the experimental strategy with the deglycosylation procedure, two experimental workflows, deglycosylation peptide analysis, and intact glycopeptide analysis, are carried out. A spectral library of glycosylation site-containing peptides in the samples is first generated by analyzing the isolated deglycosylation peptides using HCD/CID LC-MS/MS. Spectra of intact glycopeptides are selected by using diagnostic

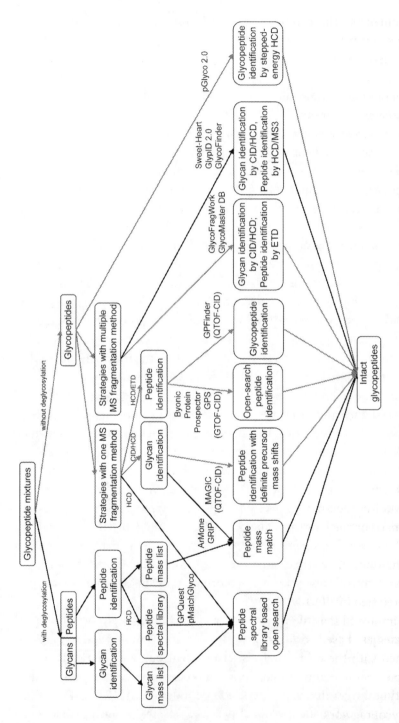

FIGURE 4.6   The overview of N-glycopeptide identification pipelines. (Translated with permission from reference Wen-Feng Zeng et al. (2016). *Progress in Biochemistry and Biophysics* 43(6):550–62.)

oxonium ions. The carrier peptide was considered a predefined mass increase according to the pre-established spectral library of glycosylation site-containing peptides. Glycan composition on a glycosylation site is determined by matching the mass difference between the precursor ion of intact glycopeptide and the deglycosylation peptide to the glycan database. There are four main software programs supporting this identification strategy: GRIP developed by Liu et al. in 2014 (Liu et al. 2014), ArMone 2.0 developed by Cheng et al. in 2014 (Cheng et al. 2014), GPQuest developed by Eshghi et al. in 2015 (Eshghi et al. 2015), and pMatchGlyco by An et al. in 2018 (An et al. 2018). The details about the software tools are described in the Chapter 5. Although GRIP, ArMone, and GPQuest can be used for high-throughput intact N-glycopeptide profiling in complex samples, the two workflows make the experimental operation tedious. Another limitation of this kind of strategy is that it is difficult both to distinguish different glycans with the same molecular weight (since the glycan moiety was deduced based on molecular weight) and to accurately match the peptide segments of the glycopeptide spectra with the peptide library.

4.3.3.2 Experimental Strategy without Extra Peptide Identification
The experimental strategy without an extra peptide identification directly identifies intact glycopeptides in MS/MS data acquired from either one type or multiple MS fragmentation methods. For example, the strategies, Glycopeptide Search (Chandler et al. 2013), and GlycoPAT (Liu et al. 2017a), which employed CID-MS/MS in spectral collection, elucidate mainly the Y-ions and partially the b/y-ions that may be detectable in CID-MS/MS spectra from highly abundant glycopeptides (e.g., standard proteins). For the strategies based on HCD-MS/MS analysis before 2015, only Y-ions were considered in HCD-MS/MS interpretation. Some strategies only used Y-ions from core-structure cleavage, while others used all possible Y-ions. From 2015 onwards, almost all pipelines include b/y ion interpretation in HCD-MS/MS. The apparent shift may be the direct result of the improved sensitivity of newer MS instruments.

Strategies based on multiple MS fragmentation methods were designed with the purpose of utilizing the combined information from multiple fragmentation due to their complementary nature. One common type is combining CID/HCD-MS/MS with ETD-MS/MS, such as GlycoFragwork (Mayampurath et al. 2014), GlycoPep Detector

(Zhu et al. 2013), GlycoMaster DB (He et al. 2014), and GlycoPAT (Liu et al. 2017a). Others include the use of AI-ETD and HCD-pd-ETD. The strategy based on SCE-HCD has been proven to be a promising method for intact glycopeptide analysis, in which informative and abundant fragment ions for both glycan and peptide of a glycopeptide could be obtained in a single spectrum. The software tool pGlyco 2.0 is useful for interpretation of this kind of strategy (Liu et al. 2017b).

4.3.4 Specific Strategies for Intact O-Glycopeptide Analysis

In principle, most of the strategies for intact N-glycopeptide analysis can be used for the O-glycopeptide analysis. However, these strategies are usually not effective for large-scale O-glycopeptide identification, and some of them are merely applied to the identification of O-glycopeptides in standard glycoproteins or very simple samples due to specific characteristics of O-glycosylation. For example, since O-GalNAcylation or mucin-type glycosylation occurs adjacently on a tryptic digested peptide, a single CID/HCD-MS/MS and even SCE-HCD-MS/MS analyses are not sufficient for O-glycopeptide analysis. ETD represents the prevailing method for O-glycosylation site localization to date. However, the approach requires peptides with high charge states so that convincing peptide fragment ions covering the O-glycosylation site can be obtained. Another challenge for O-glycopeptide analysis is the extremely low abundance of O-glycosylation. The signals of O-glycopeptides can be largely concealed by those of N-glycopeptides when analyzing O-glycopeptides without the removal of N-glycans. In addition to a range of different unbiased enrichment methodologies as described in Section 2.2 and the strategies for N-glycopeptide identification that are occasionally used for O-glycopeptide analysis as mentioned above, specific strategies for intact O-glycopeptide analysis were designed accordingly. Most of these specifically designed strategies are combined with the novelty of pretreatment and MS analysis for effective O-glycopeptide analysis.

In the early stages, some simplified strategies were developed. For example, in 2011, a strategy based on a gene-engineered cell system named "SimpleCell" was proposed for O-glycopeptide identification. In this strategy, all O-glycopeptides bear truncated O-glycans with one or two monosaccharides, which can be directly identified by a normal proteomic MS analysis and interpretation method, greatly simplifying the identification. With the development of the MS technique, more dedicated methods

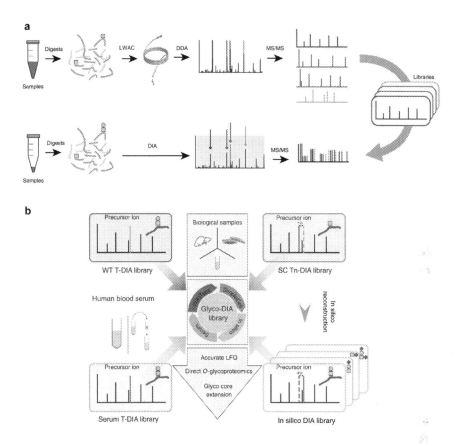

FIGURE 4.7 The Glyco-DIA method. (A) Graphic depiction of the workflow for generation of DIA glycopeptide libraries and the DIA workflow for direct glyco-proteomic analysis with Glyco-DIA libraries. (B) Overview of Glyco-DIA libraries. (Adapted with permission by reference Ye, Z. et al. (2019) *Nature Methods* 16(9): 902–10.)

for intact O-glycopeptide analysis were introduced. Recently, a data-independent acquisition (DIA) method (Figure 4.7), which relies on a spectral library partially obtained from the SimpleCell experiment, was recently developed and named "Glyco-DIA" by the same team of the "SimpleCell"; this method is useful for the characterization of glycopeptides and structures of O-glycans on a proteome-wide scale with quantification of stoichiometries (Ye and Mao 2019). This was the first attempt at large-scale O-glycopeptide identification with the DIA method and obtained impressive results.

The strategy of IsoTag as mentioned in Section 3.3 can also be used for intact O-glycopeptide. Greater than 500 intact and fully elaborated O-glycopeptides from 250 proteins across 3 human cancer cell lines were identified by IsoTag. Other effective strategies for defining the site-specific O-glycoproteome include the use of chemoenzymatic tools for selective proteolysis and detection of mucin-domain glycoproteins. Yang et al. used a commercial enzyme OpeRATOR that specifically cleaves on the N-terminal side of O-glycan-occupied Ser or Thr and coupled with ETD/ HCD-MS2 analysis for site-specific mapping of O-glycopeptides (Yang et al. 2018). Another chemoenzymatic strategy proposed by Shon et al., although not applied to MS-based O-glycopeptide identification, is an effective enzymatic toolkit for selective proteolysis, detection, and visualization of mucin-domain glycoproteins. These researchers engineered mucin-selective binding agents with retained glycoform preferences and used catalytically inactivated mucin-cleaving bacterial proteases as staining reagents to visualize the mucin-domain glycoproteins (Shon et al. 2020). An ingenious method was proposed by Di et al. for targeted analysis of both N- and O-sialylglycopeptides (N- and O-SGPs), in which, sialic acids were specifically oxidized and chemically labelled by two arginine isotopologues (Arg-$^{15}N_4$ and Arg-D^4, differs by 36 mDa). The equally mixed precursor partners, spacing tens of mDa apart, enable the direct recognition of sialylglycopeptides in MS1 level and benefit the subsequent targeted MS2 characterization, which greatly facilitated the analysis of N- and O-SGPs and resulted in the identification of 371 N-SGPs and 334 O-SGPs from human serum in a single LC-MS run in their study.

4.4 QUANTITATION OF INTACT GLYCOPEPTIDES

The quantitation of intact glycopeptides is extremely important for explorations of the roles of glycosylation in different biological and pathological stages. The quantification of intact glycopeptides at a proteome scale is still at the stage of initial development; however, MS1/MS2/MS3-based quantification methods that are normally used in proteomic quantitation have also been applied to intact glycopeptide analysis. MS1/MS2-based strategies using proteomic quantification methods, including label-free, iTRAQ and TMT isobaric labelling, for intact N-glycopeptide quantification, just combined a routine proteome quantification pipeline with intact glycopeptide enrichment and identification procedures, most of which are not specifically optimized for glycopeptide quantification. The immature experimental workflow and the lack of dedicated quantification software tools for glycopeptide quantitation resulted in reduced identification rates and impaired accuracy of quantification. SugarQuant was proposed as an integrated workflow for intact glycopeptide identification and quantification. It uses the synchronous precursor selection (SPS)-MS3 approach to generate high-resolution MS2 and MS3 fragment ions for N-glycopeptide identification, and extracts TMT reporter-ion intensities from MS3 scans for each identified N-glycopeptide-spectrum-match (GPSM) (Fang and Ji 2020). The Glyco-DIA method as mentioned above employs in silico-boosted glycopeptide libraries and enables quantitation of O-glycopeptides without enrichment for glycopeptides.

Another intact glycopeptide quantitation methods were developed for targeted quantitation. Targeted quantitation methods for glycopeptide quantitation that use targeted MS techniques, specifically multiple reaction monitoring (MRM) on triple quadruple instruments and parallel reaction monitoring (PRM) on orbitraps, have been widely adopted for quantitative evaluation of targeted glycoproteins. Currently, targeted quantitation strategies typically employ CID for the fragmentation of glycopeptides and use B-ions (e.g., m/z 204, 274, 293, 366, 513, and/or 657), Y0 or Y1 fragments as transition ions to quantify glycopeptides. Although the strategies are not suitable for large-scale glycoproteomic quantification, they were adopted as effective methods for the accurate quantitation of glycopeptides from IgG, IgA, and IgM directly from serum.

4.5 APPLICATION OF INTACT GLYCOPEPTIDE ANALYSIS FOR CLINICAL RESEARCH

Numerous studies have shown that altered glycosylation plays a key role in the pathological process during disease progression. In the past few years, MS-based glycoproteomics has emerged as a promising tool for cancer biomarker discovery-driven clinical research (Chen et al. 2019). In fact, many current disease biomarkers are glycoproteins, such as CEA for colorectal cancer, CA-125 for ovarian cancer, and AFP for hepatocellular carcinoma. However, their use remains limited due to serious technical challenges such as the complexity of the biological sample diversity of glycans and unstable changes in glycans, especially during oncogenesis (Intasqui et al. 2018; Pinho and Reis 2015). With the development of MS technologies and corresponding software, direct analysis of intact glycopeptides/glycoproteins is enabled and indisputably pushes clinical research to advanced stage (Yang et al. 2017). Intact glycopeptide analysis can provide additional site-specific glycosylation information, which allows researchers to know which glycans are attached to which glycosylation sites. Intact glycopeptide-based biomarkers hold great promise to improve the sensitivity and specificity of current protein-based biomarkers and eventually contribute to early disease diagnosis. Herein, we mainly describe the application of intact glycopeptide analysis in clinical research according to the sample source, including human body fluids and tissues (Drake et al. 2010).

4.5.1 Intact Glycopeptide Analysis in Body Liquids

Body fluid is one of the main resources for the identification of disease-related biomarkers. Explorative site-specific N- and O-glycoproteomic studies of biofluids, such as human serum, blood plasma, urine or cerebrospinal fluid (CSF), hold enormous potential to excavate candidate disease-associated glycopeptide biomarkers and better understand the implications of protein glycosylation under physiological conditions. Consequently, the identification of these intact glycopeptides in these body fluids is essential for their clinical utility.

4.5.1.1 Serum

Serum is an easily accessible sample and is one of the most extensively studied clinical samples for intact glycopeptides. For example, in 2014, Song et al. reported LC-MS/MS quantitative analysis of human blood serum glycoproteins and glycopeptides associated with oesophageal diseases by LAC- and HC-based enrichment (lectin enrichment). The author used multiple reaction monitoring quantitation and identified 13 glycopeptides by LAC enrichment and 10 glycosylation sites by HC enrichment that were significantly different among disease cohorts (oesophageal adenocarcinoma and high-grade dysplasia of metaplastic oesophageal epithelium) (Song et al. 2014). Four years later, in a large-scale study, a total of 257 glycoproteins containing 970 unique glycosylation sites and 3447 nonredundant N-linked glycopeptide variants were identified in 24 serum samples from patients with prostate cancer (Bollineni et al. 2018). Although the protein as well as the glycopeptide ratios indicated very little to no significant differences between indolent and aggressive serum prostate cancer samples, the methodology still provided clinical researchers with widely accepted and used software (Mascot), which could be easily implemented for automated glycopeptide analysis. Furthermore, in the study of intact glycopeptide analysis on liver disease, another two research groups reported their experimental results. Yang et al. reported a chemical labelling strategy to improve the ETD efficiency in MS analysis, which enables 100% peptide sequence coverage of N-glycopeptides in all subclasses of IgG (Yang et al. 2019). This comprehensive glycosylation analysis strategy for the first time allows the discrimination of IgG3 and IgG4 intact N-glycopeptides with high sequence similarity without the antibody-based pre-separation. Using this strategy, aberrant serum IgG N-glycosylation for four IgG subclasses associated with cirrhosis and hepatocellular carcinoma was revealed. Moreover, this method identifies 5-fold more intact glycopeptides from human serum than the native-ETD method, implying that the approach can also accommodate large-scale site-specific profiling of glycoproteomes. At the same time, the research group of Lubman developed a workflow that involved advanced methodologies, including the EThcD-MS/MS fragmentation method and data interpretation software, for differential analysis of the microheterogeneity of site-specific intact N-glycopeptides of serum haptoglobin between early hepatocellular carcinoma (HCC) and liver cirrhosis (Zhu and Chen 2019). In total, 93, 87, and 68 site-specific N-glycopeptides were identified in early HCC patients, patients with liver cirrhosis, and healthy controls, respectively, with high confidence. The increased variety of N-glycopeptides in patients with liver diseases compared to healthy controls was due to increased branching with hyper fucosylation and sialylation. Differential quantitation analysis showed that five site-specific N-glycopeptides on sites N184 and N241 were significantly elevated in early HCC patients compared to cirrhosis patients ($p < 0.05$) and normal controls ($p \leq 0.001$). In 2020, a novel data quantitation platform, Byos, was introduced for the characterization of the alterations in site-specific glycoforms of serum Hp in a large cohort of patients (Zhu et al. 2020). Differential quantitation analysis revealed that the levels of five N-glycopeptides at sites N184 and N241 were significantly elevated during the progression from NASH cirrhosis to HCC ($p < 0.05$). When combined with AFP, the two panels improved the sensitivity for early NASH HCC from 59% (AFP alone) to 73% while maintaining a specificity of 70%, based on the optimal cutoff. Two-dimensional (2D) scatter plots of the AFP value and N-glycopeptides showed that these N-glycopeptide markers detected 58% of AFP-negative HCC patients as distinct from cirrhosis patients. These site-specific

N-glycopeptides could serve as potential markers for the early detection of HCC in patients with NASH-related cirrhosis.

The above intact glycopeptide analysis mainly focuses on N-linked glycosylation in serum, but O-linked glycosylation also plays crucial roles in tumour growth and metastasis (Brockhausen 2006; Gill et al. 2011). In 2017, by combining HILIC tip enrichment and in silico deglycosylation method for spectral interpretation, Qin et al. identified 407 intact O-GalNAc glycopeptides from 93 glycoproteins in human serum (Qin et al. 2017). Approximately 81% of the glycopeptides contained at least one sialic acid, which could reveal the microheterogeneity of O-GalNAc glycosylation. Later that same year, an article published by Zhang et al. indicated that 21 intact O-glycopeptides in 6 glycoproteins were significantly changed between the non-CreIgAN patients and healthy human serum (ratio non-CreIgAN/healthy control <0.5 or >2), and 15 intact O-glycopeptides in 9 glycoproteins were significantly changed between serum from CreIgAN patients and healthy individuals (ratio CreIgAN/healthy control <0.5 or >2) (Zhang et al. 2018). The results underscore the potential of this strategy to discover O-glycosylation biomarkers.

In addition, sialylation is a common type of glycosylation of proteins. Aberrant sialylation of glycoproteins is closely related to many malignant diseases, and analysis of sialylation exhibits great potential in revealing the status of these diseases (Bhide and Colley 2017; Zhang et al. 2014). The research group of Mingliang Ye developed a hydrophilic smart material with switchable surface charge (histidine-bonded silica, HBS) modified with histidine that was employed for the differential analysis of glycoprotein sialylation between HCC patients and control samples (Qin et al. 2019a). Among these, 43 glycosites and 55 site-specific glycoforms exhibited significant changes in glycosite and site-specific glycoform levels, respectively. Interestingly, several glycoforms attached to the same glycosite were detected with different change tendencies.

4.5.1.2 Plasma

Similar to serum, plasma is also a major research object in clinical research. Five years ago, Hoffmann et al. developed a glycoproteomic workflow that allows the explorative nontargeted analysis of O-glycosylated human blood plasma proteins (Hoffmann et al. 2016). Site-specific glycosylation analysis of human blood plasma glycopeptides revealed exclusively mono- and disialylated core-1 mucin-type O-glycopeptides. IgG is a prominent glycoprotein in circulation that plays a key role in the humoral immune response, which is easily enriched from either serum or plasma (Wuhrer et al. 2007). Recently, Zou et al. obtained and analyzed IgG Fc N-glycopeptides in the plasma from 46 CRC patients and 67 healthy individuals using a chitosan nanomaterial (Zou et al. 2020). By analyzing IgG N-glycosylation in the plasma of CRC patients and healthy controls, the authors found that the levels of 11 N-glycopeptides such as IgG1 G0N in CRC patients significantly varied ($p<0.01$). A multivariate logistic regression model was developed using the levels of IgG Fc N-glycopeptides to distinguish CRC patients from healthy individuals; the prediction performance was good, and the average AUC of the receiver operating characteristic (ROC) curves was 0.893. In addition, the research group of Hao Yang profiled N-linked intact glycopeptides of purified IgGs from 51 prostate cancer (PCa) patients and 45 benign prostatic hyperplasia (BPH) patients (Zhang et al. 2020a). Quantitative analysis of the N-linked intact glycopeptides using pGlyco 2.0 and MaxQuant software provided quantitative information on plasma IgG subclass-specific and site-specific N-glycosylation. As a result, four aberrantly expressed N-linked intact glycopeptides across different IgG subclasses were discovered. In particular, the content of the N-glycopeptide IgG2-GP09 (EEQFNSTFR [H5N5S1]) was dramatically elevated in plasma from PCa patients, compared with that from BPH patients (PCa/BPH ratio = 5.74, p = 0.001). Furthermore, IgG2-GP09 displayed a more powerful prediction capability (auROC = 0.702) for distinguishing PCa from BPH than the clinical index t-PSA (auROC = 0.681) when used alone or in combination with other indicators (auROC = 0.853). These abnormally expressed N-linked intact glycopeptides exhibit potential for noninvasive monitoring and pre-stratification of prostate diseases.

4.5.1.3 Urine

Urine is another potentially attractive fluid for disease biomarker discovery and diagnostic tests (Pang et al. 2002). It is especially useful for the detection of lesions that are proximate to the urinary system, such as the kidney and prostate. Several publications have focused on the site-specific glycopeptide analysis of urinary glycoproteins in the last decade. First, in 2013, the research group of Moon developed a microbore hollow fibre enzyme reactor coupled to nanoflow liquid chromatography-tandem MS (nLC-ESI-MS/MS) for the online digestion or selective enrichment of glycopeptides and analysis of proteins (Kim et al. 2013). The method was applied to urine samples from patients with PCa and controls. In total, 67 N-linked glycopeptides were identified and relative differences in glyco-peptide content between patient and control samples were determined. In particular, zinc alpha-2-glycoprotein was 2.17 ± 0.44-fold more abundant in patient samples than control samples. This protein has previously been identified as a urine biomarker of prostate cancer (Katafigiotis et al. 2012). Five years later, Kawahara et al. described a glycoproteomic strategy for deep quantitative mapping of N- and O-glycoproteins in urine with the goal of investigating the diagnostic value of the glycoproteome to discrim-inate PCa from BPH (Kawahara et al. 2018). The authors accurately quan-tified 729 N-glycoproteins spanning 1310 unique N-glycosylation sites and observed 954 and 965 unique intact N- and O-glycopeptides, respec-tively, across the two disease conditions. Importantly, a panel of 56 intact N-glycopeptides perfectly discriminated PCa and BPH (ROC: AUC = 1). This study generated a panel of intact glycopeptides that has the potential for PCa detection. In another paper, Belczacka et al. carried out a larger study (Belczacka et al. 2019). The glycopeptide distribution was assessed in a set of 969 patients with five different cancer types: bladder, prostate and pancreatic cancer, cholangiocarcinoma, and renal cell carcinoma. A total of 37 intact O-glycopeptides and 23 intact N-glycopeptides were identified in the urinary profiles of 238 normal subjects. Further statisti-cal analysis revealed that three O-glycopeptides and five N-glycopeptides differed significantly in their abundance among the different cancer types compared to normal subjects. In addition to research on the urinary sys-tem, this assay is useful in thyroid cancer. Zhang et al. demonstrated that intact N-glycopeptides from the urine of healthy controls (HC), PTC, and PTC with Hashimoto's thyroiditis (PHT) were enriched by hydro-philic interaction liquid chromatography and revealed a novel indicator (ratio of fucosylated to nonfucosylated N-glycopeptide or F/NF) through desialo-N-glycopeptide analysis (Zhang et al. 2020b). These differentially expressed glycoproteins and F/NF may serve as biomarkers contributing to clinical cancer diagnostics and could be used to noninvasively improve diagnostic accuracy.

4.5.1.4 Cerebrospinal Fluid

Although CSF has also been recognized as an effective fluid for disease biomarker discovery, few researches have used CSF to analyze intact glycopeptides. Using MS-based analysis of intact O-glycopeptide from human CSF, a total of 106 O-glycosylation sites from CSF proteins were identified, and the exact attachment site for 67 of these were experimentally verified (Halim et al. 2013).

4.5.2 Intact Glycopeptides in Human Tissues

Resolving the molecular details of glycoproteome variation in different tissues and organs of the human body is critical for the understanding of human biology and disease (Uhlén et al. 2015). Intact glycopeptide analysis of disease-related tissues has provided valuable information to identify promising targets for their diagnosis, prognosis, and therapy.

Previously, based on the deglycosylation strategy, Zhu et al. mapped the N-glycosylation sites of the human liver by combining click maltose-HILIC and improved hydrazide chemistry (Zhu et al. 2014). The authors identified 14,480 N-glycopeptides matched with the consensus NXS/T motif (X≠P) from human liver, corresponding to 2210 N-glycoproteins and 4783 N-glycosylation sites. These N-glycoproteins are widely involved in different types of biological processes, such as hepatic stellate cell activation and the acute phase response of the human liver, which are all highly associated with the progression of liver diseases. Moreover, the exact N-glycosylation sites of some key-regulating proteins within different human liver physiological processes were obtained, such as E-cadherin, transforming growth factor beta receptor and 29 members of the G protein-coupled receptor family. Approximately 80% of cervical cancers occur in developing countries, especially in China (Bray et al. 2018). The well-known markers AFP and AFP-L3, which represent primarily the biomarkers for HCC monitoring and diagnosis, are continuously being investigated to improve the sensitivity and specificity of their use. The research group of Mingliang Ye mixed enzymatically de-glycosylated peptides and intact glycopeptides and analyzed them in the same LC-MS/MS run (Qin et al. 2019b). In total, 2039 and 2131 intact glycopeptides were identified from the HCC and control samples, respectively compared to the control samples, 4 of the sites (Asn2007, Asn2203, Asn2415, and Asn2693) were significantly increased (p < 0.05, ratio>2) on the relative glycosite occupancy in the HCC samples. In addition, three decreased glycoforms were high-mannose type, but all five increased glycoforms were hybrid type, including two core-fucosylated glycoforms. In addition to studies on liver cancer, in 2011, Li et al. monitored glycosylated and sialylated prostate-specific antigen (PSA) in prostate cancer and noncancer tissues by selected reaction monitoring (SRM) (Li et al. 2011). The results of this study demonstrated that PSA gly-copeptides could be detected in clinical specimens in a specific, sensitive, reproducible, and high-throughput fashion. The data suggest that differently glycosylated isoforms of glycoproteins can be quantitatively analyzed and may provide unique information for clinically relevant studies. Then, this research group introduced a chemoenzymatic method for the site-specific extraction of O-linked glycopeptides (EXoO), which has been designed to simultaneously enrich and identify O-linked glycosylation sites and define their site-specific glycans with core 1 structures with or without sialic acids. EXoO was also able to reveal significant differences in the O-linked glycoproteome of tumour and normal kidney tissues, pointing to its broader use in clinical diagnostics and therapeutics. The most striking change observed was the dramatic increase in O-linked glycans primarily in the core 1 structure Hex(1)HexNAc(1) across the 163 and 82 sites mapped in versican core protein (VCAN) and aggrecan core protein (ACAN), respectively (Yang et al. 2018).

4.6 TOWARDS MORE GENERIC AND PRECISE ANALYSIS OF INTACT GLYCOPEPTIDES IN CLINICAL RESEARCH

During the recent decades, intact glycopeptide analysis methods are advancing at a rapid rate. Many MS-based approaches for qualitative and quantitative glycopeptide analysis have sprung up, helped and will continue to help us to comprehensively profile and understand glycosylation precisely and deeply. With advances in intact glycopeptide analysis, glycoproteomics not only can generate a partial list of glycans or glycosylation sites in a given cell type or tissue but also can provide more comprehensive site-specific glycosylation information, which is undoubtedly helpful for clinical and glycobiological research. Intact glycopeptide analysis is moving towards the next stage of more precise and generic analysis. Development of more efficient approaches, such as efficient enrichment methods for O-glycopeptides, accurate site-specific glycan identification strategies, and glycopeptide quantification methods, are still needed. Elucidation of site-specific glycan's roles and its functions in physiological and pathological processes, such as virus infection, and the occurrence and development of diseases, are key future areas of investigation. We are now still only beginning to understand the remarkable significance of glycosylation, especially the site-specific glycosylation in all aspects of biology.

ACKNOWLEDGEMENTS

The authors appreciate the help in the literature research and helpful comments and suggestions from Mengxi Wu, Siyuan Kong, and Zhengzhe Huang.

REFERENCES & FURTHER READING

Aboufazeli, F., V. Kolli, E. D. Dodds. 2015. A comparison of energy-resolved vibrational activation/dissociation characteristics of protonated and sodiated high mannose N-glycopeptides. *Journal of the American Society for Mass Spectrometry* 26(4): 587–95.

Alpert, A. J. 1990. Hydrophilic-interaction chromatography for the separation of peptides, nucleic acids and other polar compounds. *Journal of Chromatography A* 499: 177–96.

An, Z., Q. Shu, H. Lv, et al. 2018. N-linked glycopeptide identification based on open mass spectral library search. *Biomed Research International* 2018: 1564136.

Belczacka, I., M. Pejchinovski, M. Krochmal, et al. 2019. Urinary glycopeptide analysis for the investigation of novel biomarkers. *Proteomics Clinical Applications* 13(3): e1800111.

Bhide, G. P., K. J. Colley. 2017. Sialylation of N-glycans: Mechanism, cellular compartmentalization and function. *Histochemistry and Cell Biology* 147(2): 149–74.

Bollineni, R. C., C. J. Koehler, R. E. Gislefoss, J. H. Anonsen, B. Thiede. 2018. Large-scale intact glycopeptide identification by mascot database search. *Scientific Reports* 8(1): 2117.

Bray, F., J. Ferlay, I. Soerjomataram, R. L. Siegel, L. A. Torre, A. Jemal. 2018. Global cancer statistics 2018: GLOBOCAN estimates of incidence and mortality worldwide for 36 cancers in 185 countries. *CA: A Cancer Journal for Clinicians* 68(6): 394–424.

Brockhausen, I. 2006. Mucin-type O-glycans in human colon and breast cancer: Glycodynamics and functions. *EMBO Reports* 7(6): 599–604.

Brodbelt, J. S. 2014. Photodissociation mass spectrometry: New tools for characterization of biological molecules. *Chemical Society Reviews* 43(8): 2757–83.

Cao, W., M. Liu, S. Kong, M. Wu, Y. Zhang, P. Yang. 2020. Recent advances in software tools for more generic and precise intact glycopeptide analysis. *Molecular and Cellular Proteomics*. doi: 10.1074/mcp.R120.002090.

Chandler, K. B., P. Pompach, R. Goldman, N. Edwards. 2013. Exploring site-specific N-glycosylation microheterogeneity of haptoglobin using glycopeptide CID tandem mass spectra and glycan database search. *Journal of Proteome Research* 12(8): 3652–66.

Chen, Z., J. Huang, L. Li. 2019. Recent advances in mass spectrometry (MS)-based glycoproteomics in complex biological samples. *Trends in Analytical Chemistry* 118: 880–92.

Cheng, K., R. Chen, D. Seebun, M. Ye, D. Figeys, H. Zou. 2014. Large-scale characterization of intact N-glycopeptides using an automated glycoproteomic method. *Journal of Proteomics* 110: 145–54.

Dai, Z., J. Zhou, S. J. Qiu, Y. K. Liu, J. Fan. 2009. Lectin-based glycoproteomics to explore and analyze hepatocellular carcinoma-related glycoprotein markers. *Electrophoresis* 30(17): 2957–66.

Darula, Z., K. F. Medzihradszky. 2018. Analysis of mammalian O-glycopeptides-we have made a good start, but there is a long way to go. *Molecular and Cellular Proteomics* 17(1): 2–17.

Di, Y., L. Zhang, Y. Zhang, et al. 2019. MdCDPM: A mass defect-based chemical-directed proteomics method for targeted analysis of intact sialylglycopeptides. *Analytical Chemistry* 91: 9986–9992.

Dongre, A., V. Wysocki. 1994. Linkage position determination of lithium-cationezed disaccharides by surface-induced dissociation tandem mass spectrometry. *Journal of Mass Spectrometry* 29: 700–2.

Drake, P. M., W. Cho, B. Li, et al. 2010. Sweetening the pot: Adding glycosylation to the biomarker discovery equation. *Clinical Chemistry* 56(2): 223–36.

Durham, M., F. E. Regnier. 2006. Targeted glycoproteomics: Serial lectin affinity chromatography in the selection of O-glycosylation sites on proteins from the human blood proteome. *Journal of Chromatography A* 1132(1–2): 165–73.

Eshghi, S. T., P. Shah, W. Yang, X. Li, H. Zhang. 2015. GPQuest: A spectral library matching algorithm for site-specific assignment of tandem mass spectra to intact N-glycopeptides. *Analytical Chemistry* 87(10): 5181–8.

Fang, P., Y. Ji. 2020. A streamlined pipeline for multiplexed quantitative site-specific N-glycoproteomics. *Nature Communications* 11(1): 5268.

Gill, D. J., H. Clausen, F. Bard. 2011. Location, location, location: New insights into O-GalNAc protein glycosylation. *Trends in Cell Biology* 21(3): 149–58.

Halim, A., U. Rüetschi, G. Larson, J. Nilsson. 2013. LC-MS/MS characterization of O-glycosylation sites and glycan structures of human cerebrospinal fluid glycoproteins. *Journal of Proteome Research* 12(2): 573–84.

Hart, G. W., R. J. Copeland. 2010. Glycomics hits the big time. *Cell* 143 (5): 672–6.

He, L., L. Xin, B. Shan, G. A. Lajoie, B. Ma. 2014. GlycoMaster DB: Software to assist the automated identification of N-linked glycopeptides by tandem mass spectrometry. *Journal of Proteome Research* 13(9):3881–95.

Hinneburg, H., K. Stavenhagen, U. Schweiger-Hufnagel, et al. 2016. The art of destruction: Optimizing collision energies in quadrupole-time of flight (Q-TOF) instruments for glycopeptide-based glycoproteomics. *Journal of the American Society for Mass Spectrometry* 27(3): 507–19.

Hoffmann, M., K. Marx, U. Reichl, M. Wuhrer, E. Rapp. 2016. Site-specific O-glycosylation analysis of human blood plasma proteins. *Molecular and Cellular Proteomics* 15(2): 624–41.

Hu, H., K. Khatri, J. Klein, N. Leymarie, J. Zaia. 2016. A review of methods for interpretation of glycopeptide tandem mass spectral data. *Glycoconjugate Journal* 33(3): 285–96.

Hu, H., K. Khatri, J. Zaia. 2017. Algorithms and design strategies towards automated glycoproteomics analysis. *Mass Spectrometry Reviews* 36(4): 475–98.

Intasqui, P., R. P. Bertolla, M. V. Sadi. 2018. Prostate cancer proteomics: Clinically useful protein biomarkers and future perspectives. *Expert Review of Proteomics* 15(1): 65–79.

Jiang, B., J. Huang, Z. Yu, et al. 2019. A multi-parallel N-glycopeptide enrichment strategy for high-throughput and in-depth mapping of the N-glycoproteome in metastatic human hepatocellular carcinoma cell lines. *Talanta* 199: 254–61.

Kaji, H., H. Saito, Y. Yamauchi, et al. 2003. Lectin affinity capture, isotope-coded tagging and mass spectrometry to identify N-linked glycoproteins. *Nature Biotechnology* 21(6): 667–72.

Katafigiotis, I., S. I. Tyritzis, K. G. Stravodimos, et al. 2012. Zinc α2-glycoprotein as a potential novel urine biomarker for the early diagnosis of prostate cancer. *BJU International* 110(11 Pt B): E688–93.

Kawahara, R., F. Ortega, L. Rosa-Fernandes, et al. 2018. Distinct urinary glycoprotein signatures in prostate cancer patients. *Oncotarget* 9(69): 33077–97.

Kim, J. Y., S. Y. Lee, S. K. Kim, S. R. Park, D. Kang, M. H. Moon. 2013. Development of an online microbore hollow fiber enzyme reactor coupled with nanoflow liquid chromatography-tandem mass spectrometry for global proteomics. *Analytical Chemistry* 85(11): 5506–13.

Li, Y., Y. Tian, T. Rezai, et al. 2011. Simultaneous analysis of glycosylated and sialylated prostate-specific antigen revealing differential distribution of glycosylated prostate-specific antigen isoforms in prostate cancer tissues. *Analytical Chemistry* 83(1): 240–5.

Li, D., Y. Chen, Z. Liu. 2015. Boronate affinity materials for separation and molecular recognition: structure, properties and applications. *Chemical Society Reviews* 44(22): 8097–123.

Liu, M., Y. Zhang, Y. Chen, et al. 2014. Efficient and accurate glycopeptide identification pipeline for high-throughput site-specific N-glycosylation analysis. *Journal of Proteome Research* 13(6): 3121–9.

Liu, G., K. Cheng, C. Y. Lo, J. Li, J. Qu, S. Neelamegham. 2017a. A comprehensive, open-source platform for mass spectrometry-based glycoproteomics data analysis. *Molecular & Cellular Proteomics* 16(11): 2032–47.

Liu, M. Q., W. F. Zeng, P. Fang, et al. 2017b. pGlyco 2.0 enables precision N-glycoproteomics with comprehensive quality control and one-step mass spectrometry for intact glycopeptide identification. *Nature Communications* 8(1): 438.

Mayampurath, A., C. Y. Yu, E. Song, J. Balan, Y. Mechref, H. Tang. 2014. Computational framework for identification of intact glycopeptides in complex samples. *Analytical Chemistry* 86(1): 453–63.

Monzo, A., G. K. Bonn, A. Guttman. 2007. Lectin-immobilization strategies for affinity purification and separation of glycoconjugates. *TrAC Trends in Analytical Chemistry* 26(5): 423–32.

Moremen, K. W., M. Tiemeyer, A. V. Nairn. 2012. Vertebrate protein glycosylation: Diversity, synthesis and function. *Nature Reviews Molecular Cell Biology* 13(7): 448–62.

Osawa, T., T. Tsuji. 1987. Fractionation and structural assessment of oligosaccharides and glycopeptides by use of immobilized lectins. *Annual Review of Biochemistry* 56: 21–42.

Pang, J. X., N. Ginanni, A. R. Dongre, S. A. Hefta, G. J. Opitek. 2002. Biomarker discovery in urine by proteomics. *Journal of Proteome Research* 1(2): 161–9.

Pinho, S. S., C. A. Reis. 2015. Glycosylation in cancer: Mechanisms and clinical implications. *Nature Reviews Cancer* 15(9): 540–55.

Prescher, J. A., D. H. Dube, C. R. Bertozzi. 2004. Chemical remodelling of cell surfaces in living animals. *Nature* 430(7002): 873–7.

Qin, H., K. Cheng, J. Zhu, J. Mao, F. Wang. 2017. Proteomics analysis of O-GalNAc glycosylation in human serum by an integrated strategy. *Analytical Chemistry* 89(3): 1469–76.

Qin, H., X. Dong, J. Mao, et al. 2019a. Highly efficient analysis of glycoprotein sialylation in human serum by simultaneous quantification of glycosites and site-specific glycoforms. *Journal of Proteome Research* 18(9): 3439–46.

Qin, H., Y. Chen, J. Mao, et al. 2019b. Proteomics analysis of site-specific glycoforms by a virtual multistage mass spectrometry method. *Analytica Chimica Acta* 1070: 60–8.

Qing, G. Y., J. Y. Yan, X. N. He, X. L. Li, X. M. Liang. 2020. Recent advances in hydrophilic interaction liquid interaction chromatography materials for glycopeptide enrichment and glycan separation. *TrAC Trends in Analytical Chemistry* 124: 115570.

Riley, N. M., A. S. Hebert, M. S. Westphall, J. J. Coon. 2019. Capturing site-specific heterogeneity with large-scale N-glycoproteome analysis. *Nature Communications* 10(1): 1311.

Shon, D. J., S. A. Malaker, K. Pedram, et al. 2020. An enzymatic toolkit for selective proteolysis, detection, and visualization of mucin-domain glycoproteins. *The Proceedings of the National Academy of Sciences U S A* 117(35): 21299–307.

Song, E., R. Zhu, Z. T. Hammoud, Y. Mechref. 2014. LC-MS/MS quantitation of esophagus disease blood serum glycoproteins by enrichment with hydrazide chemistry and lectin affinity chromatography. *Journal of Proteome Research* 13(11): 4808–20.

Sun, S., P. Shah, S. T. Eshghi, et al. 2016. Comprehensive analysis of protein glycosylation by solid-phase extraction of N-linked glycans and glycosite-containing peptides. *Nature Biotechnology* 34(1): 84–8.

Thaysen-Andersen, M., N. H. Packer. 2014. Advances in LC-MS/MS-based glycoproteomics: Getting closer to system-wide site-specific mapping of the N- and O-glycoproteome. *Biochimica et Biophysica Acta* 1844(9): 1437–52.

Thaysen-Andersen, M., N. H. Packer, B. L. Schulz. 2016. Maturing glycoproteomics technologies provide unique structural insights into the N-glycoproteome and its regulation in health and disease. *Molecular & Cellular Proteomics* 15(6): 1773–90.

Uhlén, M., L. Fagerberg, B. M. Hallström, et al. 2015. Proteomics. Tissue-based map of the human proteome. *Science* 347(6220): 1260419.

Varki, A. 2017. Biological roles of glycans. *Glycobiology* 27(1): 3–49.

Weis, W. I., K. Drickamer. 1996. Structural basis of lectin-carbohydrate recognition. *Annual Review of Biochemistry* 65: 441–73.

Woo, C. M., A. T. Iavarone, D. R. Spiciarich, K. K. Palaniappan, C. R. Bertozzi. 2015. Isotope-targeted glycoproteomics (IsoTaG): A mass-independent platform for intact N- and O-glycopeptide discovery and analysis. *Nature Methods* 12(6): 561–7.

Wu, S. W., S. Y. Liang, T. H. Pu, F. Y. Chang, K. H. Khoo. 2013. Sweet-heart - an integrated suite of enabling computational tools for automated MS2/MS3 sequencing and identification of glycopeptides. *Journal of Proteomics* 84: 1–16.

Wuhrer, M., J. C. Stam, F. E. van de Geijn, et al. 2007. Glycosylation profiling of immunoglobulin G (IgG) subclasses from human serum. *Proteomics* 7(22): 4070–81.

Xiao, H., W. Chen, J. M. Smeekens, R. Wu. 2018. An enrichment method based on synergistic and reversible covalent interactions for large-scale analysis of glycoproteins. *Nature Communications* 9(1): 1692.

Xiao, H., F. Sun, S. Suttapitugsakul, R. Wu. 2019. Global and site-specific analysis of protein glycosylation in complex biological systems with mass spectrometry. *Mass Spectrometry Reviews* 38(4-5): 356–79.

Xu, G., M. Wong, Q. Li. 2019. Unveiling the metabolic fate of monosaccharides in cell membranes with glycomic and glycoproteomic analyses. *Chemical Science* 10(29): 6992–7002.

Yang, Y., V. Franc, A. J. R. Heck. 2017. Glycoproteomics: A balance between high-throughput and in-depth analysis. *Trends in Biotechnology* 35(7): 598–609.

Yang, W., M. Ao, Y. Hu, Q. K. Li, H. Zhang. 2018. Mapping the O-glycoproteome using site-specific extraction of O-linked glycopeptides (EXoO). *Molecular Systems Biology* 14(11): e8486.

Yang, L., Z. Sun, L. Zhang, et al. 2019. Chemical labeling for fine mapping of IgG N-glycosylation by ETD-MS. *Chemical Science* 10(40): 9302–7.

Ye, Z., Y. Mao. 2019. Glyco-DIA: A method for quantitative O-glycoproteomics with in silico-boosted glycopeptide libraries. *Nature Methods* 16(9): 902–10.

You, X., H. Qin, M. Ye. 2018. Recent advances in methods for the analysis of protein o-glycosylation at proteome level. *Journal of Separation Science* 41(1): 248–61.

Zeng, W. F., M. Q. Liu, Y. Zhang, et al. 2016a. pGlyco: A pipeline for the identification of intact N-glycopeptides by using HCD- and CID-MS/MS and MS3. *Scientific Reports* 6: 25102.

Zeng, W. F., Y. Zhang, M. Q. Liu, et al. 2016b. Trends in mass spectrometry-based large-scale N-glycopeptides analysis. *Progress in Biochemistry and Biophysics* 43(6): 550–62.

Zhang, Z., Z. Sun, J. Zhu, et al. 2014. High-throughput determination of the site-specific N-sialoglycan occupancy rates by differential oxidation of glycoproteins followed with quantitative glycoproteomics analysis. *Analytical Chemistry* 86(19): 9830–7.

Zhang, Y., X. Xie, X. Zhao, et al. 2018. Systems analysis of singly and multiply O-glycosylated peptides in the human serum glycoproteome via EThcD and HCD mass spectrometry. *Journal of Proteomics* 170: 14–27.

Zhang, Y., T. Lin, Y. Zhao, et al. 2020a. Characterization of N-linked intact glycopeptide signatures of plasma IgGs from patients with prostate carcinoma and benign prostatic hyperplasia for diagnosis pre-stratification. *Analyst* 145(15): 5353–62.

Zhang, Y., W. Zhao, Y. Zhao, et al. 2020b. Comparative glycoproteomic profiling of human body fluid between healthy controls and patients with papillary thyroid carcinoma. *Journal of Proteome Research* 19(7): 2539–52.

Zhu, Z., D. Hua, D. F. Clark, E. P. Go, H. Desaire. 2013. GlycoPep detector: A tool for assigning mass spectrometry data of N-linked glycopeptides on the basis of their electron transfer dissociation spectra. *Analytical Chemistry* 85(10): 5023–32.

Zhu, J., Z. Sun, K. Cheng, et al. 2014. Comprehensive mapping of protein N-glycosylation in human liver by combining hydrophilic interaction chromatography and hydrazide chemistry. *Journal of Proteome Research* 13(3): 1713–21.

Zhu, Z., H. Desaire. 2015. Carbohydrates on proteins: Site-specific glycosylation analysis by mass spectrometry. *Annual Review of Analytical Chemistry (Palo Alto Calif)* 8: 463–83.

Zhu, J., Z. Chen. 2019. Differential quantitative determination of site-specific intact N-glycopeptides in serum haptoglobin between hepatocellular carcinoma and cirrhosis using LC-EThcD-MS/MS. *Journal of Proteome Research* 18(1): 359–71.

Zhu, J., J. Huang, J. Zhang. 2020. Glycopeptide biomarkers in serum haptoglobin for hepatocellular carcinoma detection in patients with nonalcoholic steatohepatitis. *Journal of Proteome Research* 19(8): 3452–66.

Zielinska, D. F., F. Gnad, J. R. Wiśniewski, M. Mann. 2010. Precision mapping of an in vivo N-glycoproteome reveals rigid topological and sequence constraints. *Cell* 141(5): 897–907.

Zou, Y., J. Hu, J. Jie, et al. 2020. Comprehensive analysis of human IgG Fc N-glycopeptides and construction of a screening model for colorectal cancer. *Journal of Proteomics* 213: 103616.

CHAPTER 5

Analytical Software and Databases in N-Glycoproteomics

Suideng Qin and Zhixin Tian

CONTENTS

5.1 INTRODUCTION

N-glycosylation, being one of the most abundant protein post-translational modifications (PTMs) (Ohtsubo and Marth 2006), plays various important roles in biological regulation, cellular recognition (Soliman et al. 2017), mediation of cellular functions (Ohtsubo and Marth 2006) and pathological processes such as carcinogenesis (Pinho and Reis 2015). N-glycosylation

DOI: 10.1201/9781003185833-5

is the attachment of N-glycans to peptide N-X-S/T sequons (where X can be any amino acid residue except P) and demonstrates both micro-heterogeneity and macro-heterogeneity (Marino et al. 2010). N-glycoproteins are universal biomarkers of several diseases, especially cancers (Berretta et al. 2017). With the advancements in enrichment, isotopic labelling, LC-MS/MS and bioinformatics, large-scale analysis of N-glycosylation has been feasible with three distinct approaches at different molecular level, i.e., N-glycan (N-glycomics), N-glycosite-containing peptides (shot-gun proteomics) and intact N-glycopeptides (bottom-up N-glycoproteomics). Each of these approaches has its advantages and is complementary to each other; bottom-up N-glycoproteomics has the advantage of simultaneously providing site- and structure-specific qualitative and quantitative information (Dallas et al. 2013).

The identification and characterization of an intact N-glycopeptide from its MS/MS spectrum is a multi-step process including spectrum filtering, database search and identification of the peptide backbone (amino acid sequence and N-glycosite), database search and identification of the N-glycan moiety (monosaccharide composition, sequence and linkages), and final identification of the intact N-glycopeptide with false discovery rate (FDR) control. N-glycan databases, search engines and annotation tools are the three key components in the process, which has recently been well summarized in some journal review articles (Abrahams et al. 2020).

Common databases for N-glycoproteomics (Abrahams et al. 2020; Lisacek et al. 2017) include but not limited to *GlycoStore* (Zhao et al. 2018), *UniProt* (Farriol-Mathis et al. 2004), *UniCarbKB* (Campbell et al. 2014), *SugarBind Database* (Shakhsheer et al. 2013), *Glyco3D* (Perez et al. 2015), *UniCarb-DB* (Hayes et al. 2011), *GlycomeDB* (Ranzinger et al. 2008), *UniPep* (Zhang et al. 2006).

Auxiliary tools have also been developed for annotation of MS spectra, N-glycan compositions and putative structures as well as peptide sequences reflected in the spectral patterns. *GlycopeptideGraphMS* (Choo et al. 2019) and *GlycoDomainViewer* (Joshi et al. 2018) are such tools.

Search engines play the major role of identification and quantitation of intact N-glycopeptides, and several tools have been successfully developed and widely adopted. These tools include commercial Byonic, research-grade pGlyco, GPQuest, GPSeeker, I-GPA, Mascot, SugarQb, MAGIC. This chapter mainly provides a brief overview and introduction of these tools. For each tool, the major search algorithm and procedure, method of FDR control and recent applications are described.

Moreover, there are also a small number of software tools for O-glycoproteomics presented in order to offer a comprehensive overview of analytical software for glycoproteomics.

5.2 N-GLYCOPROTEOMIC SEARCH ENGINES AND TOOLS

5.2.1 Byonic

Byonic (Bern et al. 2007), first reported in 2007, is the only commercial intact N-glycopeptide search engine that introduces an algorithm of "look-up peaks" to promote the identification of MS/MS spectra generated from multiple types of fragmentation techniques. The most recent edition supports low-energy collision-induced dissociation (CID), beam-type CID, time of flight (TOF)-TOF and electron transfer dissociation (ETD)/electron capture dissociation (ECD) (Bern et al. 2012). "Look-up peaks" refers to ion peaks that include specific features, usually b- or y-ion peaks.

To conduct identification, Byonic makes use of both parent ions and look-up peaks. Its identification process contains three basic steps including: computation of look-up peaks, search of the database to find candidates and scoring of the candidates.

For Step 1, look-up peaks are identified and computed in pairs, before which intensities of peaks including isotopic peaks need to be manipulated. Each pair of peaks is recorded as (i, j), which corresponds to the pair of peaks in the spectrum that has a mass difference of one amino acid residue. The complementary pair (M-j, M-i) would be calculated when parent ion mass is known and all pairs of look-up peaks will then be sorted, selected and filtered.

For Step 2, Byonic uses either peak pairs or individual peaks to find candidates in the database by computing the number of matches with look-up peaks at mass falling within the window of parent ion mass and then selects the candidate peptides depending on their tryptic termini and number of matches.

For Step 3, the spectra of candidate peptides selected from the database are processed with the same procedures as described in Step 1 (manipulate peaks intensities in order to reflect peak ranks). Then, theoretical peaks and expected intensities are generated for each candidate peptide, and finally, it gives out the score which is a sum of positive points for theoretical peaks found in the modified spectrum and negative points for those not found.

In addition, Byonic provides a menu of several variable modifications, especially the internal table of most possible N- and O-glycans, and it assumes that there will be at most one variable modification per peptide in order to precisely identify peptides with PTMs, which enables accurate identification of intact N-glycopeptides. Moreover, Byonic can remove all the peaks of the top-scoring peptides and generate "knockout spectrum" to conduct second identification, so that it can handle multiplexed spectra.

Byonic utilizes decoy protein database to compute FDR. To lower the false positive rate, Byonic offers an editable threshold where look-up peak pairs with inadequate total intensities are regarded as noise and rejected, so that Byonic is able to enhance spectrum quality and improve the accuracy of spectrum identification. False positive and false negative rate, sensitivity as well as comparison with other search engines are evaluated with mouse blood plasma spiked with low concentrations of recombinant human proteins.

As Byonic performs peptide search with considerable precision, sensitivity and validity, it has been widely applied as normalized search engine for site-specific intact N-glycopeptide identification. Byonic has been applied to the glycoproteomic analysis of multiple systems such as RNase B and immunoglobulin gamma molecules (IgGs) from the HepG2 cell line (Yu et al. 2017), human hepatocyte lysate (Greer et al. 2018), GFR-TKI-sensitive PC9 and GFR-TKI-resistant PC9-IR cells from non-small cell lung cancer (Waniwan et al. 2018), human urinary proteins (Pap et al. 2018), PKM2 knockout breast cancer cells (Chen et al. 2018), novel biomarkers in urinary intact N-glycopeptide (Belczacka et al. 2019), human serum haptoglobin between early hepatocellular carcinoma and liver cirrhosis (Zhu et al. 2019).

For instance, in 2017, Zixiang Yu et al. (Yu et al. 2017) analyzed glycopeptides from the HepG2 cell line using nano-RPLC-MS/MS and Byonic; 4514 intact N-glycopeptides were identified with 1011 unique peptide backbones and 947 N-glycosites.

In 2018, Yu-Ju Chen et al. (Waniwan et al. 2018) used Orbitrap higher-energy collisional dissociation (HCD)-CID-MS/MS to analyze site-specific glycoproteins in non-small cell lung cancer; 2290 and 2967 unique intact N-glycopeptides were identified with Byonic score≥100 from EGFR-TKI-sensitive PC9 and EGFR-TKI-resistant PC9-IR cells, respectively. Lingjun

Li et al. (Chen et al. 2018) utilized hydrophilic interaction liquid chromatography (HILIC), multi-lectin affinity enrichment and electron-transfer/higher-energy collision dissociation (EThcD) fragmentation technique to analyze N-glycoproteins in PANC1 cells; 1067 intact N-glycopeptides from 205 N-glycoproteins were identified, revealing 311 N-glycosites and 88 N-glycan compositions; moreover, their approach was further applied to N-glycosylation alteration in PKM2 knockout cells.

In 2019, David M. Lubman et al. (Zhu et al. 2019) conducted characterization and differential quantitation research on intact N-glycopeptides from serum haptoglobin among early hepatocellular carcinoma (HCC), liver cirrhosis and healthy controls through LC-EThcD-MS/MS and Byonic/Biologic; 93, 87 and 68 intact N-glycopeptides were identified, respectively.

5.2.2 pGlyco 2.0

pGlyco 2.0 (Liu et al. 2017) is a dedicated search engine designed for high-throughput stepped collision energies (SCE) HCD-MS/MS acquisition method which is based on optimized collision parameters to perform precise large-scale N-glycoproteome identification and analysis of complex samples with comprehensive quality control (FDR evaluation of N-glycan moieties, peptide backbones and intact N-glycopeptides). The entire workflow of the intact N-glycopeptide interpretation and FDR evaluation (Figure 5.1) using pGlyco 2.0 mainly includes three steps: firstly, glycoproteomic sample is analyzed by SCE-HCD-MS/MS and processed spectra are generated; secondly, these spectra are searched against N-glycan and protein databases (DBs) to perform identification; finally, the FDR of N-glycan, peptide backbone and intact N-glycopeptide are estimated comprehensively to generate labelled spectra.

In the aforementioned MS acquisition, SCE under 20-30-40% were used in HCD to generate the most informative fragment ions of both the N-glycan moiety and the peptide backbone from an intact N-glycopeptide in a single spectrum; pGlyco 2.0 utilizes this comprehensive information to conduct integrated search of each spectrum.

To perform identification of intact N-glycopeptides, each combined spectrum is sequentially scored against the glycome database and proteome database to select N-glycan candidates and peptide backbone candidates. Then, intact N-glycopeptide candidates are generated by combining candidate N-glycans and peptide backbones. In detail, the identification

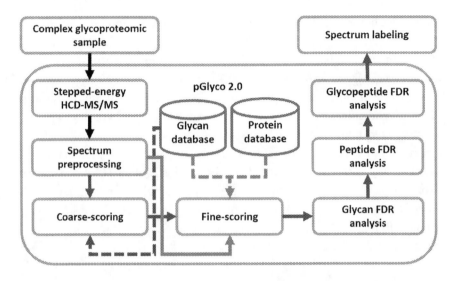

FIGURE 5.1 Procedure of identifying and interpreting intact N-glycopeptides by pGlyco 2.0 (Liu et al. 2017).

procedure contains coarse-scoring and fine-scoring. In coarse-scoring, for each N-glycan in the glycome database, the mass of the associated peptide backbone could be calculated as the precursor mass minus the N-glycan moiety mass and then all the possible masses of Y ions can be computed. Each N-glycan is scored depending on the number of matched Y ions in the corresponding spectrum with the demand for at least two matched trimannosyl core ions. In fine-scoring, for each valid candidate N-glycan refined by coarse-scoring, the corresponding candidate peptides are searched by peptide DB search engine pFind (Wang et al. 2007) utilizing the mass of the peptide backbone; finally, both N-glycan and peptide candidates are combined to generate intact N-glycopeptide candidates. The scoring section contains N-glycan scoring (calculated by internal algorithm using the matched peaks, the number of matched trimannosyl core ions and mass errors), peptide scoring (similar to N-glycan scoring) and total intact N-glycopeptide scoring (calculated as the weighted sum of the former two scores).

For comprehensive quality control, FDR of matches to N-glycans, peptides and intact N-glycopeptides should all be computed. Since both the N-glycan moiety and the peptide backbone searching procedure could result in false identification, pGlyco 2.0 estimates intact N-glycopeptide

FDR by using internal algorithm to combine the false identification of both N-glycan and peptide, which can be obtained by the target-decoy approach. To benchmark FDR control, researchers established two search engine-independent methods: isotope-based FDR and entrapment-based FDR. In the former, equal amount of yeast proteins, metabolically labelled by ^{13}C and ^{15}N, together with unlabelled proteins were used as sample, while a combinatorial DB of yeast glycome and proteome database was searched as the target DB; the reported N-glycopeptide spectrum matches (GPSMs) that failed to match with the unlabelled and corresponding labelled pair in the MS1 scan were considered as false positives. In the latter, mouse-only glycome and proteome DB were searched as the entrapment DB and any GPSMs reported from the entrapment database were judged as false positives. Then, the false positive from these two methods were computed and utilized to validate the FDR reported by the search engine. As a result, pGlyco 2.0 achieved high precision in FDR estimation and this validation workflow can be applied to compare the FDR validity of different search engines. All in all, pGlyco 2.0 enables FDR estimation of all three types of matches at one quality control program with high confidence.

pGlyco 2.0 was applied to the N-glycosylation analysis of IgGs in prostate carcinoma and benign prostatic hyperplasia to conduct diagnosis pre-stratification (Zhang et al. 2020).

In the study of behaviours and performances of universal intact N-glycopeptide enrichment methods using boronic acid chemistry, zwitterionic chromatography-hydrophilic interaction liquid chromatography (ZIC-HILIC) and porous graphitic carbon (PGC) (Xue et al. 2018), intact N-glycopeptides from mouse liver were enriched by different enrichment methods and interpreted by pGlyco 2.0. Boronic acid chemistry tended to capture N-glycopeptides with high-mannose N-glycans, ZIC-HILIC did not appear to have obvious preference and PGC was not applicable to the enrichment of N-glycopeptides with long amino acid sequence.

In the analysis of alpha-1-antitrypsin N-glycosylation which contains core- and antennary-fucosylated N-glycosites using LC-MS/MS (Yin et al. 2018), three N-glycosites were detected with pGlyco search: Asn70 contains biantennary N-glycans without fucosylation; both Asn107 and Asn271 have core- and antennary-fucosylation; Asn107 contains bi-, tri- and tetra- antennary N-glycans while Asn271 only contains bi- and tri-antennary N-glycans.

5.2.3 GPQuest

GPQuest (Eshghi et al. 2015) utilizes spectral library matching to conduct site-specific identification and characterization of intact N-glycopeptide in complex samples analyzed by LC-MS/MS (HCD).

The entire workflow of intact N-glycopeptide identification using GPQuest can be coarsely depicted in the following four steps: (1) establishment of the experimental spectral library (ESL), (2) selection of intact N-glycopeptide spectra, (3) matching the intact N-glycopeptide spectra with the ESL to identify the N-glycosite-containing peptides and (4) assignment of the N-glycans at each N-glycosite (Figure 5.2).

For Step 1, 90% of the sample is first enriched using solid-phase extraction of N-linked glycosite-containing peptide (SPEG) method and treated with PNGase F to isolate N-glycosite-containing peptides. Then, these peptides are analyzed by HCD LC-MS/MS and the spectral results are interpreted by SEQUEST in the Proteome Discoverer software, resulting in peptides (including sequence and glycosite information) matching for each spectrum. In order to build ESL, b- and y-fragment ions that range from singly to triply charged are generated for each identified peptide and then the MS/MS spectra of each identified peptide are searched for the presence of the above-mentioned ions. Finally, the identified

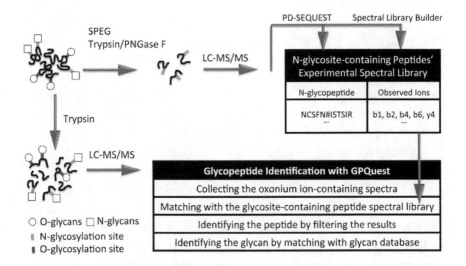

FIGURE 5.2 Workflow of identifying intact N-glycopeptides using GPQuest (Eshghi et al. 2015).

N-glycosite-containing peptides (which are used as target peptides) and their corresponding observed b- and y-ions are compiled to generate ESL, whereas those peptides with less than four observed b- and y-ions are removed.

For Step 2, the remaining 10% of the sample is treated with trypsin and analyzed by HCD LC-MS/MS. Then, the generated raw files are processed and the presence of signature oxonium ions (which belong to fragmented N-glycan free monosaccharides or disaccharides) are searched to distinguish the glycopeptides spectra from common peptide spectra. The selected oxonium-ion-containing spectra (containing at least two identified oxonium ions with masses corresponding to predetermined saccharides) are considered as intact N-glycopeptide spectra.

For Step 3, since the patterns of b- and y-ions are similar between the spectra of fragmented intact glycopeptides and their de-glycosylated peptide counterparts, the percentage of the shared experimental b- and y-ions in each MS/MS spectrum of intact glycopeptides and each ESL peptide is calculated to conduct the peptide matching process. Then, for each target peptide in the ESL, a series of intact peptide ions (total nine types, including different charge states and different attached partial N-glycans) are generated and the aforementioned peptide matches are filtered based on the number of the generated ions which are observed in the intact N-glycopeptide spectra (the minimum required number of the generated ions for each peptide is determined by the length of the peptide). The matching results are then further refined by removing matches whose overlap with the ESL do not exceed an FDR-based threshold. In addition, peptide b- and y-ions with partial N-glycan attachments can be utilized to assign the remaining peaks to intact N-glycopeptide ions. Finally, the peptide backbone of each intact N-glycopeptide can be identified based on the final matching results.

For Step 4, the mass of the N-glycan moiety of each intact N-glycopeptide is calculated as the corresponding corrected precursor mass minus the mass of the identified peptide backbone. The N-glycan composition is then determined by searching against an N-glycan DB to match with the calculated N-glycan mass. It is possible that the calculated N-glycan masses are missing from the N-glycan DB. Therefore, spectral library matching strategy allows highlighting the spectra containing uncharacterized N-glycans and investigates them to figure out the structure of those N-glycans. All in all, GPQuest is able to conduct identification and characterization of intact N-glycopeptides analyzed by HCD LC-MS/MS using a pre-built ESL and N-glycan DB.

To control the false positives reported in the process of matching, GPQuest utilizes target-decoy approach to estimate FDR. The decoy DB is generated by combining the amino acids of all identified SPEG-enriched N-glycosite-containing peptides, randomly rearranging them and dividing them into decoy peptide sequences with the same length as the target peptides in the target DB. Any peptide spectrum match (PSM) reported from the decoy DB is considered as a false positive. Moreover, a threshold of 40% for the peptide matching refinement in Step 3 will lead to an FDR of approximately 1%.

GPQuest was applied to the site-specific analysis of atypical N-glycosites on albumin and α-1B-glycoprotein (serum glycoproteins) (Sun et al. 2018). Half of the identified N-glycosites were attached with at least two N-glycans, and two novel atypical N-X-V glycosites were identified.

In the reanalysis of global proteomic and phospho-proteomic data from breast cancer tissues with GPQuest 2.0 (Hu et al. 2018), 6,724 intact N-glycopeptides from 617 glycoproteins were identified from two breast cancer xenograft tissues. Also, the proteomic and phospho-proteomic data set of 108 human breast cancer tissues generated by Clinical Proteomic Analysis Consortium (CPTAC) were reanalyzed, resulting in identification of total 11,292 N-glycopeptides corresponding to 1,731 N-glycosites from 883 human glycoproteins.

In the glycoproteomic analysis of Chinese hamster ovary (CHO) cell (Yang et al. 2018), 10,338 N-glycosite-containing intact glycopeptides (IGPs) which represented 1,162 unique glycosites in 530 glycoproteins were identified. In addition, 71 atypical N-linked IGPs on 18 atypical N-X-C sequons were observed.

5.2.4 Integrated GlycoProteome Analyzer (I-GPA)

I-GPA is an N-glycoproteome search engine (Park et al. 2016) developed for automated identification, label-free quantitation and quantitative comparison of site-specific N-glycopeptides analyzed by nRPLC HCD- and CID-MS/MS. I-GPA is composed of a GPA database (GPA-DB), an identification algorithm (id-GPA), a quantitation component (q-GPA) and a comparison component (c-GPA), enabling high-throughput mapping of complex N-glycoproteomes. GPA-DB for each experimental sample is

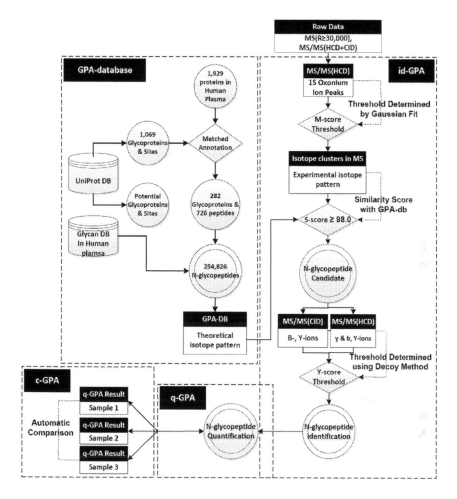

FIGURE 5.3   Search process and related algorithms of I-GPA (Park et al. 2016).

constructed by GPA-DB-Builder software and the information for masses and relative intensities in isotope pattern is included. The graphic search algorithms and identification procedure of I-GPA can be viewed below (Figure 5.3).

The procedure of automated identification of intact N-glycopeptides using id-GPA contains three steps: (1) selection of intact N-glycopeptides from HCD-MS/MS spectra utilizing specific oxonium ions; (2) selection of candidate intact N-glycopeptides by matching the isotope pattern to the intact N-glycopeptides in the GPA-DB; and (3) identification of intact N-glycopeptides from CID- and HCD-MS/MS fragment ions.

For Step 1, a total of 15 specific oxonium ions that are differently weighted depending on their frequency of appearance are utilized to select intact N-glycopeptide spectra from HCD-MS/MS spectra due to their prominent sensitivity as markers of intact N-glycopeptide fragmentation (usually singly protonated mono- and oligo-saccharide ions generated from intact N-glycopeptide fragmentation) in HCD-MS/MS spectra. In order to compute the frequency of the appearance of those oxonium ions, M-score is set up as a criterion and calculated using a specific equation that contains the number of excepted and matched oxonium ions, the peak intensity of matched ions as well as the highest intensity of peak less than 700 Da. Calculated M-score within a M-score threshold is capable of revealing and picking out MS/MS spectra that include intact N-glycopeptide markers.

For Step 2, after the MS/MS spectra of intact N-glycopeptide were selected, the isotope patterns of their corresponding precursor ions are collected and searched against the GPA-DB to find the best match, where isotope patterns of molecular ions between theoretical and experimental data are compared and the similarity is estimated by calculating S-score as a criterion. S-score is calculated using a specific equation containing the m/z values and intensities of the theoretical and experimental isotopic peaks. Finally, the best match is determined by setting a S-score threshold, so that intact N-glycopeptide candidates including peptides and glycoforms can be selected.

For Step 3, since both CID and HCD fragmentations are performed with peculiarity such as the abundant Y- & B-ions generated in CID-MS/MS spectra and the appearance of y- & b-ions (revealing oxonium ions and the amino acid sequences of the peptide backbone) in HCD-MS/MS spectra, id-GPA utilizes both CID and HCD spectra as well as their characteristics to conduct identification of intact N-glycopeptides. Experimental CID- and HCD-MS/MS spectra are first compared with the theoretical spectra of the intact N-glycopeptide candidates. Then, the matches of CID and HCD fragment peaks between the theoretical and experimental spectra are calculated as CID_{match} and HCD_{match}, respectively, using two specific equations with the same form containing the number of peaks, the intensity of the highest peak, the intensity of the matched peak and the individual peak intensity. CID_{match} and HCD_{match} are then combined to select the true identifications of intact N-glycopeptide as many as possible and a CID_{match}:HCD_{match} ratio of 7:3 leads to the highest area under curve (AUC) value. Therefore, a Y-score criterion defined as $Y_{score} = 0.7*CID_{match} + 0.3*HCD_{match}$ is set up to score and determine the final intact N-glycopeptide identifications with an appropriate Y-score threshold.

To estimate FDR in the identification procedures and conduct quality control, a decoy approach is utilized in id-GPA. Once the S-scoring is complete, the decoy B- and Y-ion candidates are generated by changing the amino acid sequences of the peptide portions and the number of Hex, HexNAc, Fuc, NeuAc for glycoforms. Then, a decoy MS/MS database, which contains peptides with reversed sequences and N-glycan moieties with exchanged numbers of monosaccharide composition, is constructed. This decoy DB is then searched and used to calculate Y-score, receiving a Y-score distribution that can promote to distinguish false positive. As a result, selecting true identifications of N-glycopeptides using the Y-score ≥69.5 can reach an estimated FDR≤0.9% and is manually validated in the experiment section.

In addition, I-GPA enables users to conduct high-speed label-free quantitation and comparative analysis of multiple samples by using q-GPA and c-GPA algorithms. An algorithm named top three-isotopes quantification (TIQ) is developed to perform quantitation, in which the combined intensity of the top three isotopic peaks at three highest MS spectral points is used as the essential reference. 3TIQ-based quantitation is then implemented using three data points with the highest values, and quantification data of intact N-glycopeptides in each sample is generated after which comparative analysis is carried out. For multiple samples quantified by q-GPA, all the q-GPA data are combined to compile a list of intact N-glycopeptides with information including isotope pattern, abundance and retention time. Then, abundances of intact N-glycopeptides quantified in all samples by q-GPA are obtained without any additional processing, while obtaining abundances of those not identified (determined from isotope pattern and retention time in the compiled list of intact N-glycopeptides) in some samples needs extra processing. The corresponding N-glycopeptides are determined using S-score approach and the abundances of these intact N-glycopeptides are obtained using 3TIQ method. Finally, the similarity and difference of intact N-glycopeptide quantitation can be analyzed by comparing the obtained abundance.

I-GPA has been applied to characterization of differential N-glycosylation at the intact N-glycopeptide level in multiple systems. For HCC plasma (Park et al. 2016), 619 intact N-glycopeptides belonging to 123 N-glycoproteins were identified and 598 were quantified.

In the analysis of site-specific microheterogeneity of N-glycans of biosimilar fusion protein VEGFR-IgG (Hahm et al. 2019), five N-glycosites were identified and microheterogeneity of N-glycans for each N-glycosite are well exhibited.

5.2.5 Mascot

Search engine Mascot (Bollineni et al. 2018) has long been adopted to the identification of peptides and interpretation of peptide sequences for proteomics. In recent years, Mascot has been improved and adapted to conduct qualitative and quantitative analysis of intact N-glycopeptides using CID or HCD fragmentation techniques at lower normalized collision energy (NCE) values in large-scale study of glycoproteomics.

Annotating the sequences of peptide portions and structures of N-glycan moieties in MS2 spectra is achieved with three steps of: (1) creating unique one-letter codes to represent monosaccharides; (2) linearizing monosaccharide sequences of N-glycans; and (3) creating customized intact N-glycopeptide database.

For Step 1, the Latin alphabet is used to form the one-letter codes for monosaccharides that appears in N-glycan compositions. Since 20 of the alphabet letters are used to code the standard amino acid residues and letter B, X, Z are hard-coded, letter J, O, U are the only choices for the one-letter codes. For the assignment of the three letters to different monosaccharides, J is assigned to Gal and Man, O to GlcNAc and GalNAc and U to Neu5Ac, while Fucose is defined as variable modifications on O (GlcNAc or GalNAc).

For Step 2, after coding the monosaccharides, the N-glycan sequence can be represented by a combination of letter J, O, U. To represent N-glycan branches and antennae, linear sequences expressed by one-letter codes should cover the maximum possible intense peaks in MS2 spectrum (usually oxonium ions and Y-ions) and reveal as much information of N-glycan sequences as possible. The linear sequence of intact N-glycopeptide is generated by combining the linear sequences of both the N-glycan moiety and peptide backbone together with N-glycosite information. Furthermore, the Y-ions (glycosidic cleavage) can be transformed into y-ions (peptide cleavage) by attaching the linear N-glycan sequences to the corresponding peptide N-termini.

For Step 3, following an in-silico digestion, the customized database is created using the combination of linear N-glycans and tryptic peptide sequences containing N-X-S/T/C motifs (or S/T residues if the search of O-glycopeptides is needed).

Briefly, searching against the customized DB using Mascot requires a series of Y-, b- and y-ions; Mascot annotates the intense peaks in MS2 spectrum to these ions. The linear N-glycan sequence is determined by computing the mass difference between the most intense peaks of different Y_n ions (peptide bound glycosidic cleavage ions) as well as the mass difference between the precursor ion and the Y0 ion. Then, the N-glycan structure can be deduced using the predetermined assignment of linear N-glycan sequences to N-glycan structures. For peptide sequencing, detected intense peaks of y-ions are needed and the peptide sequence is determined by computing the precursor mass and the mass difference between different peaks of y_n ions (peptide cleavage type ions) in MS2 spectrum. However, HCD or CID fragmentation at a single NCE value may not provide adequate information for N-glycan and peptide sequencing. In order to enrich the fragmentation information of the N-glycan moiety and the peptide backbone, stepped lower NCE values (15, 25 and 35) are adopted to HCD fragmentation. As a result, intensities of different y-ion peaks vary at diverse NCE values. By utilizing this feature, Mascot is able to make use of intense y-ion peaks at different NCE values to create a composite MS2 spectrum and annotate almost the entire peptide part of the intact N-glycopeptide sequence. Moreover, the label-free relative quantification is performed using Mascot Distiller and the quantitative differences can be found by segmenting intact N-glycopeptides according to the N-glycan structures.

Mascot has been applied to the large-scale comparative serum study of the sialylated N-glycoproteome from prostate cancer patients and healthy individuals (Bollineni et al. 2018). Zwitterionic chromatography-hydrophilic interaction liquid chromatography solid phase extraction (ZIC-HILIC SPE) was used to desalt tryptic peptides and enrich intact N-glycopeptides, and LC-MS with stepped NCE HCD fragmentation was used to analyze samples. A total of 3447 intact N-glycopeptides from 970 N-glycosites on 257 glycoproteins were identified.

In the site-specific analysis of N-glycoproteome heterogeneity in complex mouse brain-derived membrane protein mixtures (Parker et al. 2013) using PGC-LC-MS/MS with CID and HCD fragmentation, 863 intact N-glycopeptides from 276 N-glycosites on 161 N-glycoproteins were identified.

5.2.6 SugarQb

SugarQb is a glycoproteomic platform (Stadlmann et al. 2017) developed as an insertion for MS/MS search engines (such as Mascot, SEQUEST-HT, X! Tandem) to perform additional identification of intact N-glycopeptides. Accompanied with a customized glycoproteomic workflow, SugarQb also enables interpretation of N-glycosylation in proteomic systems and quantification of intact N-glycopeptides.

The search procedure of SugarQb is composed of three steps: extraction and pre-processing of MS/MS spectra, selection of MS/MS spectra of intact N-glycopeptides and identification of peptide sequences and N-glycan structures.

For Step 1, MS/MS spectra are extracted from the raw data using Spectrum Exporter Node of Proteome Discoverer 1.4 and are pre-processed with precursor mass correction, charge deconvolution and de-isotoping (Savitski et al. 2010). The signals of multiply charged fragment ions are transformed into the corresponding singly charged m/z values. Fragment ions of intact peptide with partly and gradually detached N-glycans (Y-ions) are transferred to the high m/z region, while peptide backbone fragment ions with less abundance (b- and y-ions) remain in the low m/z region. In the pre-processed MS/MS spectrum, the high and low m/z regions are separated by the intense Y1 ion (peptide + HexNAc).

For Step 2, N-glycosylation-related reporter ions (m/z 126.05496, 138.05496, 144.06552) are used for selection of MS/MS spectra of intact N-glycopeptides. Then, the presence and abundances of these reporter ions are searched and computed to acquire a G-score. The G-score here is calculated as the reciprocal value of the negative logarithm of the summed value of normalized reporter ion intensities divided by their respective rank in the spectrum. A G-score threshold is then set up to promote selection, above which the given MS/MS spectrum is regarded as an intact N-glycopeptide spectrum.

For Step 3, the reporter ions are first removed from the selected spectra. Then, the intact N-glycopeptide spectra are searched for the presence of Y1 ions with a mass tolerance by iteratively subtracting all masses in N-glycan DB minus 203.0794 a.m.u from the mass of the respective precursor ion. Only singly charged ions are considered to conduct fragment-ion peak matching; and once a potential Y1 ion is detected, the MS/MS spectrum is cloned, with its original precursor mass being set to the mass of Y1 ion to generate a processed MS/MS data. All the processed data are then searched against the Uniprot proteome database (containing forward and reverse database) using Mascot or X! Tandem search engine to obtain the amino acid sequences of peptide backbones, with a single HexNAc moiety considered as a variable modification to any asparagine residue that might be the potential N-glycosite. The reported PSMs are filtered manually based on the criteria of search-engine rank Top1, peptide length>6 and containing at least one HexNAc residue as modification. For the identification of N-glycan structures, the monosaccharide compositions of N-glycans are first acquired from the process of identifying potential Y1 ions in which an N-glycan DB is utilized to promote identification of Y1 ions as well as to determine the compositions of N-glycans. The N-glycan structures are then deduced using known biosynthetic rules of N-glycosylation. Finally, the in-house developed ptmRS module is adopted for N-glycosite localization (Taus et al. 2011).

SugarQb has been applied to the comparative analysis of human and mouse embryonic stem cells (hESCs and mESCs) (Stadlmann et al. 2017). With IP-HILIC enrichment and nano-LC-ESI-MS/MS (HCD), 3380 intact glycopeptides from 508 mESC N-glycoproteins and 1106 intact N-glycopeptides from 576 hESC N-glycoproteins were identified; 237 were found to be shared revealing conserved and specific N-glycan modifications at a proteome scale.

SugarQb was migrated to Proteome Discoverer 2.1 in conjunction with Mascot and MS Amanda search engines, and was applied to the analysis of PNGase F-resistant N-glycopeptides from mESCs (Stadlmann et al. 2018). For the PNGase F-treated samples, 365 and 242 intact N-glycopeptides were identified using Mascot and MS Amanda, respectively. A total of 985 intact N-glycopeptide analyzed by nanoRPLC-ESI-MS/MS upon PNGase F-treatment were comparatively quantified.

5.2.7 MAGIC

MS-based Automated Glycopeptide IdentifiCation (MAGIC) (Lynn et al. 2015) is a software platform developed to perform identification of peptide sequences and N-glycan compositions of intact N-glycopeptides from high-throughput beam-type CID MS2 datasets in conjunction with Mascot and X! Tandem search engines. A search algorithm named Trident is implemented in MAGIC; and *in silico* spectra are generated during the search process to conduct peptide backbone sequencing. For the determination of N-glycan compositions, MAGIC constructs the fragmentation path of the N-glycan moieties.

The identification workflow of MAGIC is composed of the four steps of filtering intact N-glycopeptide MS2 spectra, detection of Y1 ion using Trident algorithm, sequencing of peptide backbones with *in silico* spectra and final determination of N-glycan compositions.

For Step 1, MAGIC detects the abundant B ions in low *m/z* region as signal ions to select potential intact N-glycopeptide spectra. B ions including HexNAc$^+$ (*m/z* 204.08) and HexHexNAc$^+$ (*m/z* 366.14) are stored in a user-editable B ion list, and the spectra with B ions no less than a user-specified threshold will be considered as intact N-glycopeptide spectra.

For Step 2, the dissociation of the trimannosyl core structure of intact N-glycopeptides in CID fragmentation will generate seven peaks (Y0-NH$_3$, Y0, Y1, Y2, Y3, Y4, Y5), where Y0 stands for the peptide backbone ion, Y1 stands for [peptide + GlcNAc] ion and Y5 stands for [peptide + 2GlcNAc + 3Man] ion. Based on this fragmentation feature, the Trident algorithm utilizes specific triplet patterns to accurately identify the Y1 ion, in which two unique triplets of peaks with *m/z* differences of two sequential GlcNAc and *m/z* differences of a GlcNAc subsequent to a NH$_3$ (corresponding to [Y0, Y1, Y2] and [Y0-NH$_3$, Y0, Y1] unique triplet peaks, respectively) are regarded as the minimum required triplet peaks to detect the Y1 ion. Once the two above-mentioned triplet patterns are detected, the corresponding Y1 ion is identified. Upon determination of the Y1 ion, other Y ions are detected by searching the *m/z* region above Y0 ion for successive peaks with *m/z* differences corresponding to any monosaccharide. Moreover, Trident enables detection of modified triplet patterns such as a water loss or sodium adduct to identify the Y1 ion.

For Step 3, MAGIC first processes MS2 spectra by removing all peaks of B and Y ions and reassigning the precursor mass to the Y0-ion mass, subsequently generating *in silico* peptide MS2 spectra. By accomplishing this process, the interference with detection of peptide b and y ions (with low relative abundance) caused by dominant B and Y ions is eliminated and the precursor masses of the corresponding peptide backbones are corrected. The generated *in silico* spectra are then searched using peptide sequence-based search engine such as Mascot and X! Tandem to interpret the peptide sequences of intact N-glycopeptides.

For Step 4, a look-up table listing all the potential N-glycan compositions (together with their corresponding masses) is first constructed by collecting exhaustive combinations of 29 different monosaccharides. Then, the mass of the target unidentified N-glycan is calculated as the corresponding intact N-glycopeptide precursor mass minus the mass of the peptide backbone. The N-glycan composition is then determined by matching the N-glycan mass to the masses listed in the look-up table, during which all the detected B and Y ions as well as the related biosynthetic rules are taken into consideration as additional criteria. Each reported N-glycan composition match is then scored to reveal the best composition match and MAGIC outputs the final results including peptide sequence and N-glycan compositions with the corresponding scores.

For confidence evaluation, FDR of peptide backbone sequencing is controlled by Mascot or X! Tandem; whereas FDR of other search procedures in MAGIC (such as identification of Y1 ion) is estimated by adopting *Escherichia·coli* dataset as a negative control. *E.·coli* does not have any N-glycosylation, and no intact N-glycopeptide was identified in *E.·coli* indicating FDR of 0%. In the benchmark study of N-glycosylation in HeLa cells with tryptic digestion and ZIC-HILIC enrichment, 36 N-glycopeptides with 22 monosaccharide compositions with identified from 26 N-glycoproteins.

5.2.8 GPSeeker

GPSeeker is an intact N-glycopeptide search engine (Xiao and Tian 2019) capable of performing automated high-throughput search of complex intact N-glycopeptide samples analyzed by LC-MS/MS with stepped normalized collision energy (NCE), revealing comprehensive structural information of the peptide backbones (amino acid sequences and N-glycosites) and the N-glycan moieties (monosaccharide compositions, sequences and glycosidic linkages). Besides qualitative identification, GPSeeker also comes together with the quantitative search module GPSeekerQuan for relative quantitation using the isotopic peak abundance of the isotopically labelled paired precursor ions in the MS spectra.

To perform qualitative identification and characterization of intact N-glycopeptides, GPSeeker is developed as an integration of intact protein and peptide database search engine ProteinGoggle (Xiao et al. 2017) and N-glycan database search engine GlySeeker (Xiao et al. 2018); both search engines adopt the search algorithm of isotopic mass-to-charge ratio and envelope fingerprinting (iMEF) (Li and Tian 2013) which is a combination of isotopic m/z fingerprinting (iMF) and isotopic envelope fingerprinting (iEF).

The protein-level search process using iMEF algorithm in ProteinGoggle can be depicted as follows: the raw experimental MS/MS spectra and

LC-MS/MS dataset are first converted into the mzXML format, while a customized theoretical DB containing the theoretical isotopic envelopes (iEs) of both precursor and product ions of every proteoform from different charge states are generated. The iMF is then used to search for precursor ions with a series of search parameters and m/z tolerance in MS level to obtain preliminary protein candidates. In this search procedure, the m/z value of most abundance isotopic peak is searched (together with its left and right adjacent isotopic peaks if the preliminary candidate is not found) in the database within the precursor ion isolation window. When a preliminary candidate is found, the searched m/z values are excluded. This search is repeated until the most abundance isotopic peak in the remaining MS spectrum is below a preset relative abundance threshold. The iEF of each preliminary candidate is then carried out to screen protein candidate(s), where the experimental isotopic peaks are then fingerprinted with all the corresponding theoretical ones. Matched product ions of every protein candidate in the experimental MS/MS spectra are searched in the same manner. Protein spectrum matches (PrSMs) are then filtered with the percentage of matched product ions (PMPs) and the PTM score. PrSMs from both the target and decoy searches are combined, and final protein IDs are obtained with FDR control and duplicates removal.

The N-glycan-level search process using iMEF algorithm in GlySeeker can be described as follows: in order to facilitate convenient computer handling of the N-glycan product ions, a universal nomenclature is developed with the combination of a capital English letter, a Roman number and an Arabic number to name the type of fragment ions, the branch position from top to bottom and the fragment ion position along the branch. The product ions on the core structure "NNM" are numbered only once and the position order is in accord with the first antenna. A comprehensive theoretical N-glycan DB is first created using the retro-synthetic strategy (Kronewitter et al. 2009) and the reported biosynthetic rules (Goldberg et al. 2005; Kornfeld and Kornfeld 1985; Lowe and Marth 2003). The iEs of both the precursor and product ions are computed and stored. The creation of theoretical human N-glycan DB is herein used as an example to describe the creation process in detail. The biggest structures of the three major types of N-glycans (high-mannose, complex and hybrid) are first built. To gain the exhaustive list of theoretical N-glycans, monosaccharides are "detached" one by one until the core structure is the only portion left. A specific N-glycan structure is adopted as the original structure for high-mannose N-glycans; six putative antennae (including Y(31F)41L31Y(31F)41L21F etc.) are adopted for the construction of complex N-glycans with a limit of at most 4 antennae and 15 monosaccharides; and for hybrid N-glycans, the two antennae from the top (numbered I and II) are identical to those of high-mannose while antenna III or IV can be any of the aforementioned six putative antennae. N-glycans are searched and identified by GlySeeker in the same manner as proteins by ProteinGoggle described above.

The search and identification procedure of intact N-glycopeptide in GPSeeker (Figure 5.4) consists of the following five steps: (1) creation of the customized intact N-glycopeptide DB from the corresponding proteome DB and N-glycome DB; (2) search of peptide backbones using Y1 ions, i.e., peptide spectrum matches (PSMs); (3) search of N-glycan moieties using the mass difference between the precursor intact N-glycopeptide and the peptide backbone, i.e., N-glycan spectrum matches (GSMs); (4) search of intact N-glycopeptide spectrum matches from the combination of PSMs and GSMs; (5) combination of target and decoy GPSMs to give final intact N-glycopeptide IDs with FDR control and duplicate removal.

For Step 1, the proteome DB is parsed to generate the N-sequon DB with the criteria of N-X-S/T/C motifs, enzyme rules, missed cleavages allowed and minimum sequence length required. Then, the customized

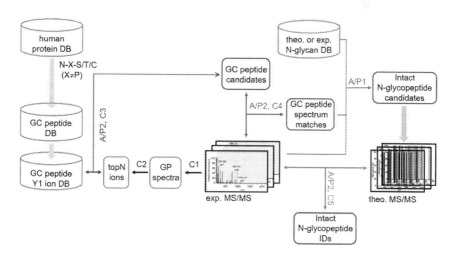

FIGURE 5.4 Search procedure of GPSeeker for intact N-glycopeptides identification and characterization (Xiao and Tian 2019).

DB of N-sequon Y1 ions is created with user-accessible options (including dissociation methods, MS spectrum m/z acquisition range, dynamic/static modifications etc.), which contains the theoretical isotopic envelopes of both the precursor and fragment ions.

For Step 2, the MS/MS spectra of intact N-glycopeptides are selected utilizing the characteristic oxonium ions with the first set of criteria (C1, within the given m/z tolerance).

For Step 3, top N fragment ions in MS/MS spectra are selected for the search of N-sequon Y1 ions using the iMEF algorithm with the second set of search parameters (including isotopic abundance cutoff, etc.) and the third set of criteria (C3; m/z value>500; observation of Y0, Y1, any other ions from Y2, Y0-NH_3, $^{0,2}X1$, etc.). Then, the candidate N-glycosite-containing peptides are determined. Theoretical fragment ions of each candidate N-glycosite-containing peptide are then generated and searched against the corresponding experimental MS/MS spectrum using iMEF algorithm with the second set of search parameters. An additional criterion of percentage of matching product ions (PMPs) is used for selection of PSMs. For unambiguous localization of N-glycosites, glyco-brackets (G-brackets) is adopted; G-bracket is defined as the number of pairs of GlcNAc-containing b and y fragment ions with each pair being able to independently unambiguously localized the N-glycosite.

For Step 4, a comprehensive customized DB of intact N-glycopeptides is firstly built by adopting an exhaustive combination of N-glycosite-containing spectrum matches with N-glycans from the customized N-glycan DB. For each PSM, iMEF of the isolated precursor ion in the corresponding MS spectra is searched against the above-mentioned comprehensive DB with the first set of search parameters to generate the corresponding candidate intact N-glycopeptides (containing putative N-glycan). Then, for each candidate intact N-glycopeptide, iMEF of the theoretical N-glycan fragment ions is utilized to obtain N-glycan spectrum matches (GSMs). The GSMs are searched against the corresponding experimental fragment ions with the second set of search parameters and the fifth set of criteria (C5) to gain the final GSMs. The uniqueness of N-glycan structure is characterized with glycoform score which is defined as the number of matched structure-diagnostic fragment ions, each of which can independently distinguish the structure with all the other putative structures with the same monosaccharide composition. For a given precursor ion, intact N-glycopeptide spectrum matches are formed with combination of PSMs and GSMs.

For Step 5, the P scores of PSMs (p-score) and GSMs (g-score) are scored respectively using the previously reported approaches from ProteinGoggle for proteins and GlySeeker for N-glycans. Then, a comprehensive pg-score is computed for GPSMs based on the above two scores.

To control the spectrum-level FDR, GPSeeker adopts target-decoy approach. The decoy databases of N-glycosite-containing peptides and N-glycans are generated by reversing the amino acid sequences and randomly adding 1-30 Da to the experimental MS/MS spectra, respectively. Then, the decoy databases are searched with the same search parameters and criteria as the target databases. Finally, both target and decoy GPSMs are combined and ranked with decreasing pg-score and a cutoff score is chosen to achieve the spectrum-level FDR≤1%.

In addition, relative quantitation of the identified intact N-glycopeptides is conducted by GPSeekerQuan. Both light- and heavy-labels are used for DB search and the paired isotopic envelopes are search in the MS spectra for each ID; the relative ratio is calculated using the peak abundance of the top three isotopic peaks in each isotopic envelope.

GPSeeker has been applied to the characterization of differential N-glycosylation in a series of cancer cells at the intact N-glycopeptide level with isotopic labelling and RPLC-nanoESI-MS/MS (HCD with stepped NCEs).

For HCC HepG2 cells relative to LO2 cells (Xiao and Tian 2019), 5405 and 1081 intact N-glycopeptides were identified and quantified respectively and the different differential expressions were detected on 183 out of the 231 quantified N-glycosites. Both sialic acid linkage-specific (Figure 5.5) and fucose position-specific (Figure 5.6) differential expression was observed for HepG2 cells, which highlights the need of structure-specific identification and quantitation.

For breast MCF-7 cancer cells relative to epithelial MCF-10A cells (Xue et al. 2020), 581 intact N-glycopeptide IDs were identified from five technical replicates and 56 DEGPs were quantified from 23 N-glycosites on 19 intact N-glycoproteins. For the 19 intact N-glycoproteins, 14, 6 and 1 were observed with uniform down, up and both down- and upregulation, respectively, with one or more DEGPs each.

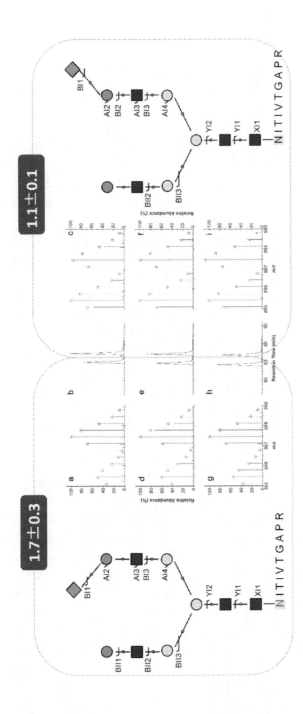

FIGURE 5.5 Sialic acid linkage-specific differential expression observed in HepG2 cells relative to LO2 cells. The α2,3 and α2,6 linkage isomers from intact N-glycopeptides NITIVTGAPR_N4H5F0S1 (N-glycosite N265 of integrin α-3, P26006, ITA3_HUMAN) was observed with upregulation (1.7±0.3) (a/d/g, left LC peaks in b/e/h) vs. No-Change (1.1±0.1) (c/f/i, right LC peaks in b/e/h), respectively. Modified from Figures 4 and 5 of reference (Xiao and Tian 2019).

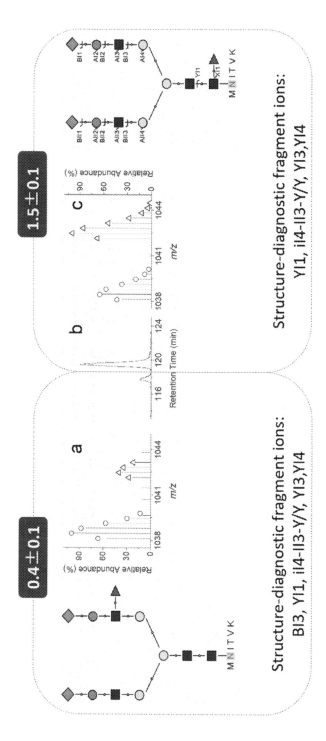

FIGURE 5.6 Fucose position-specific differential expression observed in HepG2 cells relative to LO2 cells. The branch and core isomers of intact N-glycopeptides MNITVK_N4H5F1S2 (N-glycosite N107 of integrin α-3, P26006, ITA3_HUMAN) were observed with downregulation (0.4±0.1) (a, left LC peaks in b) *vs.* upregulation (1.5±0.1) (c, right LC peaks in b), respectively. Modified from supplementary Figures S9 of reference (Xiao and Tian 2019).

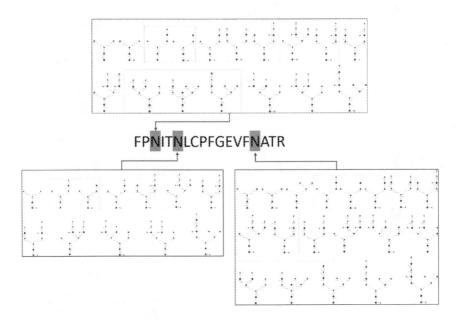

FIGURE 5.7   Site- and structure-specific N-glycosylation of a vaccine targeting the RBD of the S protein of SARS-CoV-2.

For MCF-7/ADR cancer stem cells (CSCs) relative to MCF-7/ADR cells (Wang et al. 2020), 1336 intact N-glycopeptides from the combination of 169 putative N-glycan linkages and 301 unique peptide backbones were identified corresponding with 289 glycoproteins and 305 N-glycosites. 72 cell-surface DEGPs were observed in which 8 and 64 were down and up regulated, respectively.

Most recently, GPSeeker was adopted for structure-specific N-glycosylation characterization of a vaccine targeting the RBD of the S protein of SARS-CoV-2 (Yang et al. 2020) (Figure 5.7); 17 and 19 distinct N-glycan structures were identified on N-glycosite N331 and N343, respectively; 12 N-glycan structures including two pairs of sequence isomers were also observed on the less frequent N-X-C sequon (Figure 5.8).

Apart from the aforementioned eight tools, many other search engines and software platforms developed and reported in recent years (Abrahams et al. 2020) also perform accurate identification and valid characterization of N-glycosylation at the N-glycan or intact N-glycopeptide level (Table 5.1).

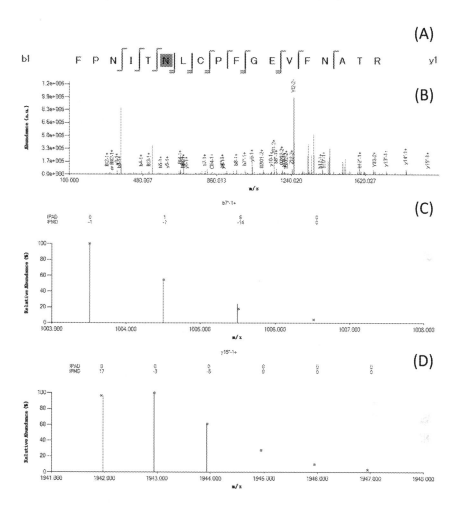

FIGURE 5.8 Site-specific N-glycosylation characterization on N-glycosite N334 with the N-X-C sequon. The N-glycosites can be confirmed with GlcNAc-containing site-determining paired bracket fragment ions of b ions (b6, b7, b8, b10, b12) with y ions (y13, y14, y15). (A), graphical fragmentation map of the peptide backbone with the matched b/y ions, red and green means without and with GlcNAc; (B), annotated MS/MS spectrum; (C) and (D), isotopic envelope fingerprinting maps of b7*-1+ and y15*-1+ (*, GlcNAc-containing).

TABLE 5.1   List of Software Tools for N-Glycomics and N-Glycoproteomics

Name	Major Functions and Website Links (if Available)
Byonic	Performing sequence-based identification of glycopeptides utilizing "lookup peaks" and assigning monosaccharide compositions (Bern et al., 2007); http://www.proteinmetrics.com/products/byonic/
GlycoMod	Determine monosaccharide compositions and predict possible structures (Cooper et al., 2001); http://www.expasy.ch/tools/glycomod
Glycoforest	Interpreting N-glycan sequences from MS/MS spectra using de novo algorithm (Horlacher et al., 2017); https://glycoforest.expasy.org
GlycoWorkbench	Evaluating all structures by matching the theoretical list against experimental MS data with different fragmentation types (Ceroni et al., 2008); http://www.eurocarbdb.org/applications/ms-tools
GlycoProteomics Analysis Toolbox (GlycoPAT)	Cataloguing site-specific N-glycosylation with FDR estimation and scoring algorithm from MS/MS spectra (Neelamegham et al., 2015); http://www.VirtualGlycome.org/glycopat
GlycoPep Detector (GPD)	Identifying intact N-glycopeptides from ETD data adopting interference filtration and candidate scoring (Zhu et al., 2013); http://glycopro.chem.ku.edu/ZZKHome.php
GlycoPep Grader (GPG)	Determining compositions of intact N-glycopeptide with user-inputted peptide sequences and theoretical N-glycans (Woodin et al., 2012); http://glycopro.chem.ku.edu/GPGHome.php
GlycoPeptideSearch	Identifying and annotating intact N-glycopeptides from MS/MS spectra (Pompach et al., 2012); https://bitbucket.org/glycoaddict/glycopeptidegraphms
I-GPA	Site-specific identification and comparative quantification of intact N-glycopeptides (Park et al., 2016); https://www.igpa.kr
GlycoMaster DB	Analyzing combinatorial HCD/ETD spectra to identify peptide sequences and putative N-glycan structures (He et al., 2014) http://www-novo.cs.uwaterloo.ca:8080/GlycoMasterDB
SimGlycan	Characterizing N-glycan structures using database search and confidence scoring (Apte and Meitei, 2010); http://www.premierbiosoft.com/glycan
GlySeeker	Conducting structure-specific N-glycan database search utilizing isotopic envelope fingerprinting and pre-built comprehensive theoretical N-glycans (75,888 entries) with sequence and linkage structures which are represented in both pseudo 2D and one-line text formats; combination of capital English letter, Roman number and Arabic number is adopted for uniform nomenclature of fragment ions; spectrum-level FDR control was achieved with target-decoy search; structure-diagnostic fragment ions are used for unambiguous structure assignment and the graphical structures of final IDs can be output in a high-throughput manner in a single PDF file (Xiao et al., 2018)
GPQuest	Site-specific identification of intact N-glycopeptides using spectral library matching algorithm, together with automated and high-throughput assignment of HCD-MS/MS spectra (Eshghi et al., 2015)

TABLE 5.1 (*Continued*)   List of Software Tools for N-Glycomics and N-Glycoproteomics

Name	Major Functions and Website Links (if Available)
GlycoMine	Systematic *in silico* identification and type-specific prediction of glycosites using random forest algorithm (Li et al., 2015) http://www.structbioinfor.org/Lab/GlycoMine
Mascot	Sequence-based search engine that enables identification of intact N-glycopeptides using linearized N-glycan sequences and automated annotation of MS/MS spectra (Bollineni et al., 2018)
SugarQb	Software platform developed in-house for identification and comparative analysis of intact N-glycopeptides analyzed by nano-LC-ESI-MS/MS (Stadlmann et al., 2017)
MAGIC	Mass spectrometry-based Automated Glycopeptide Identification platform Software platform capable of identifying intact N-glycopeptides from beam-type CID spectra with specified filtering technique and search algorithms (Lynn et al., 2015) http://ms.iis.sinica.edu.tw/COmics/Software_MAGIC.html
MultiGlycan-ESI	Automatically annotating and label-free quantitation of N-glycans from ESI-LC-MS/MS spectra (Hu et al., 2015) http://darwin.informatics.indiana.edu/MultiGlycan/
pGlyco	Identification and quantitation of intact N-glycopeptides from high-throughput SCE-HCD spectra with comprehensive quality control (Liu et al., 2017) http://pfind.ict.ac.cn/download/pGlyco/pGlyco2-stable.zip
Sweet-Heart	Performing identification of N-glycopeptides from ion trap-based LC-MS/MS data by implementing a machine learning algorithm (Liang et al., 2014)
NGlycPred	Predicting the N-glycan occupancy at the sequons of N-X-S/T utilizing both the structure and residue patterns of the corresponding glycoprotein (Chuang et al., 2012) http://exon.niaid.nih.gov/nglycpred
Glycosylation Prediction Program (GPP)	Accurately predicting glycosites using random forest algorithm and pairwise patterns (Hamby and Hirst, 2008) http://comp.chem.nottingham.ac.uk/glyco
GPSeeker	Automated site- and structure-specific identification and quantitation of intact N-glycopeptides from LC-MS/MS (HCD with stepped NCEs) analysis of complex N-glycoproteomes with enrichment and isotopic labelling. Spectrum-level FDR is controlled with target-decoy search. Unambiguous N-glycosite localization and structure assignment are achieved with site-determining and structure-diagnostic fragment ions, respectively; Dynamic modification (such as acetylation, methylation, phosphorylation) other than N-glycosylation and static labelling (such as alkylation and isotopic labelling) are fully supported. Relative quantitation is achieved with isotopic peak abundance of the paired precursor ions in MS spectra. (Xiao and Tian, 2019)

5.3 SOFTWARE TOOLS FOR O-GLYCOPROTEOMICS

Apart from N-glycosylation, O-glycosylation is another type of glycosylation which commonly presents with prominent structural diversity and is regarded as one of the prevalent PTMs. Resembling the biological and physiological functions of N-glycosylation, O-glycosylation shows close relevance to cellular metabolism (Grammel and Hang 2013), pathological processes, cellular recognition, etc. Thus, O-glycosylation is frequently considered to be potential biomarker of various diseases (such as cancer, cardiovascular diseases) (Dove 2001; Pinho and Reis 2015), indicating the great importance of O-glycosylation in many biological processes. In recent years, LC-MS/MS pipeline has been adopted to conduct site-specific large-scale identification and characterization of O-glycopeptides. However, identification and characterization of O-glycoproteomes are much more challenging than those of N-glycoproteomes. Firstly, O-glycosylation is much less abundant than N-glycosylation which leads to weak signals in MS/MS spectra and big difficulty of detecting diagnostic ions; secondly, O-glycan is much easier to break during fragmentation, causing the loss of structural features in MS/MS spectra; thirdly, O-glycosites do not present on uniform sequons (such as N-X-S/T motifs for N-glycosites), which increases the difficulty of O-glycosite prediction and assignment; and finally O-glycosylation exhibits prominent microheterogeneity and macroheterogeneity. So far, only few search engines have been developed for specific database search and identification from O-glycoproteomic data; and the major ones are Byonic (Belczacka et al. 2019; Pap et al. 2018; Zhang et al. 2018), SEQuestHT (SugarQb) (Stadlmann et al. 2017), GPQuest (Eshghi et al. 2015), Mascot (Bollineni et al. 2018) and O-GlycoProteome Analyzer (O-GPA) (Park et al. 2020). In addition, only several of tens of common O-glycan compositions are normally supported per search.

Wide range (tens to thousands) of O-glycan compositions have been reported in different studies, and a comprehensive theoretical O-glycome DB based on current biosynthetic rules is urgently needed. However, thousands of O-glycans on multiple putative amino acids (S/T/Y) is a huge search space, which needs an efficient solution.

5.4 CONCLUSIONS

Current intact N-glycopeptide search engines have made high-throughput site- and structure-specific identification and quantitation of N-glycosylation possible in LC-MS/MS based bottom-up N-glycoproteomics, which will undoubtedly accelerate mapping of human N-glycosylation landscape and discovering of more effective glycoprotein diagnostic markers and drug targets. More complex structure features of N-glycosylation, such as post-glycosylational modifications, remain as a big challenge and need further research. Large-cohort clinical study will be the major future effort of the field which will surely bring exciting outcome.

REFERENCES

Abrahams, J. L., G. Taherzadeh, G. Jarvas, A. Guttman, Y. Q. Zhou, M. P. Campbell. 2020. Recent advances in glycoinformatic platforms for glycomics and glycoproteomics. *Current Opinion in Structural Biology* 62: 56–69.

Apte, A., N. S. Meitei. 2010. Bioinformatics in glycomics: Glycan characterization with mass spectrometric data using SimGlycan. *Methods in Molecular Biology* 600: 269–81.

Belczacka, I., M. Pejchinovski, M. Krochmal, et al. 2019. Urinary glycopeptide analysis for the investigation of novel biomarkers. *Proteomics Clinical Applications* 13: 1800111.

Bern, M., Y. H. Cai, D. Goldberg. 2007. Lookup peaks: A hybrid of de novo sequencing and database search for protein identification by tandem mass spectrometry. *Analytical Chemistry* 79: 1393–400.

Bern, M., Y. J. Kil, C. Becker. 2012. Byonic: Advanced peptide and protein identification software. *Current Protocols in Bioinformatics* 40: 13.20.1–13.20.14.

Berretta, M., C. Cavaliere, L. Alessandrini, et al. 2017. Serum and tissue markers in hepatocellular carcinoma and cholangiocarcinoma: Clinical and prognostic implications. *Oncotarget* 8: 14192–220.

Bollineni, R. C., C. J. Koehler, R. E. Gislefoss, J. H. Anonsen, B. Thiede. 2018. Large-scale intact glycopeptide identification by mascot database search. *Scientific Reports* 8: 2117.

Campbell, M. P., R. Peterson, J. Mariethoz, et al. 2014. UniCarbKB: Building a knowledge platform for glycoproteomics. *Nucleic Acids Research* 42: D215–21.

Ceroni, A., K. Maass, H. Geyer, R. Geyer, A. Dell, S. M. Haslam. 2008. GlycoWorkbench: A tool for the computer-assisted annotation of mass spectra of glycans. *Journal of Proteome Research* 7: 1650–9.

Chen, Z. W., Q. Yu, L. Hao, et al. 2018. Site-specific characterization and quantitation of N-glycopeptides in PKM2 knockout breast cancer cells using DiLeu isobaric tags enabled by electron-transfer/higher-energy collision dissociation (EThcD). *Analyst* 143: 2508–19.

Choo, M. S., C. Wan, P. M. Rudd, T. Nguyen-Khuong. 2019. GlycopeptideGraphMS: Improved glycopeptide detection and identification by exploiting graph theoretical patterns in mass and retention time. *Analytical Chemistry* 91: 7236–44.

Chuang, G. Y., J. C. Boyington, M. G. Joyce, et al. 2012. Computational prediction of N-linked glycosylation incorporating structural properties and patterns. *Bioinformatics* 28: 2249–55.

Cooper, C. A., E. Gasteiger, N. H. Packer. 2001. GlycoMod: A software tool for determining glycosylation compositions from mass spectrometric data. *Proteomics* 1: 340–9.

Dallas, D. C., W. F. Martin, S. Hua, J. B. German. 2013. Automated glycopeptide analysis–review of current state and future directions. *Brief Bioinform* 14: 361–74.

Dove, A. 2001. The bittersweet promise of glycobiology. *Nature Biotechnology* 19: 913–7.

Eshghi, S. T., P. Shah, W. M. Yang, X. D. Li, H. Zhang. 2015. GPQuest: A spectral library matching algorithm for site-specific assignment of tandem mass spectra to intact N-glycopeptides. *Analytical Chemistry* 87: 5181–8.

Farriol-Mathis, N., J. S. Garavelli, B. Boeckmann, et al. 2004. Annotation of post-translational modifications in the SWISS-PROT knowledge base. *Proteomics* 4: 1537–50.

Goldberg, D., M. Sutton-Smith, J. Paulson, A. Dell. 2005. Automatic annotation of matrix-assisted laser desorption/ionization N-glycan spectra. *Proteomics* 5: 865–75.

Grammel, M., H. C. Hang. 2013. Chemical reporters for biological discovery. *Nature Chemical Biology* 9: 475–84.

Greer, S. M., M. Bern, C. Becker, J. S. Brodbelt. 2018. Extending proteome coverage by combining MS/MS methods and a modified bioinformatics platform adapted for database searching of positive and negative polarity 193 nm ultraviolet photodissociation mass spectra. *Journal of Proteome Research* 17: 1340–7.

Hahm, Y. H., J. Y. Lee, Y. H. Ahn. 2019. Investigation of site-specific differences in glycan microheterogeneity by N-glycopeptide mapping of VEGFR-IgG fusion protein. *Molecules* 24: 3924.

Hamby, S. E., J. D. Hirst. 2008. Prediction of glycosylation sites using random forests. *BMC Bioinformatics* 9: 500.

Hayes, C. A., N. G. Karlsson, W. B. Struwe, et al. 2011. UniCarb-DB: A database resource for glycomic discovery. *Bioinformatics* 27: 1343–4.

He, L., L. Xin, B. Z. Shan, G. A. Lajoie, B. Ma. 2014. GlycoMaster DB: Software to assist the automated identification of N-linked glycopeptides by tandem mass spectrometry. *Journal of Proteome Research* 13: 3881–95.

Horlacher, O., C. Jin, D. Alocci, et al. 2017. Glycoforest 1.0. *Analytical Chemistry* 89(20): 10932–40.

Hu, Y. L., S. Y. Zhou, C. Y. Yu, H. X. Tang, Y. Mechref. 2015. Automated annotation and quantitation of glycans by liquid chromatography/electrospray ionization mass spectrometric analysis using the MultiGlycan-ESI computational tool. *Rapid Communications in Mass Spectrometry* 29: 135–42.

Hu, Y. W., P. Shah, D. J. Clark, M. H. Ao, H. Zhang. 2018. Reanalysis of global proteomic and phosphoproteomic data identified a large number of glycopeptides. *Analytical Chemistry* 90: 8065–71.

Joshi, H. J., A. Jorgensen, K. T. Schjoldager, et al. 2018. GlycoDomainViewer: A bioinformatics tool for contextual exploration of glycoproteomes. *Glycobiology* 28: 131–6.

Kornfeld, R., S. Kornfeld. 1985. Assembly of asparagine-linked oligosaccharides. *Annual Review of Biochemistry* 54: 631–64.

Kronewitter, S. R., H. J. An, M. L. de Leoz, C. B. Lebrilla, S. Miyamoto, G. S. Leiserowitz. 2009. The development of retrosynthetic glycan libraries to profile and classify the human serum N-linked glycome. *Proteomics* 9: 2986–94.

Li, L., Z. X. Tian. 2013. Interpreting raw biological mass spectra using isotopic mass-to-charge ratio and envelope fingerprinting. *Rapid Communications in Mass Spectrometry* 27: 1267–77.

Li, F., C. Li, M. Wang, et al. 2015. GlycoMine: a machine learning-based approach for predicting N-, C- and O-linked glycosylation in the human proteome. *Bioinformatics* 31: 1411–9.

Liang, S. Y., S. W. Wu, T. H. Pu, F. Y. Chang, K. H. Khoo. 2014. An adaptive workflow coupled with random Forest algorithm to identify intact N-glycopeptides detected from mass spectrometry. *Bioinformatics* 30: 1908–16.

Lisacek, F., J. Mariethoz, D. Alocci, et al. 2017. Databases and associated tools for glycomics and glycoproteomics. *Methods in Molecular Biology* 1503: 235–64.

Liu, M. Q., W. F. Zeng, P. Fang, et al. 2017. pGlyco 2.0 enables precision N-glycoproteomics with comprehensive quality control and one-step mass spectrometry for intact glycopeptide identification. *Nature Communications* 8: 438.

Lowe, J. B., J. D. Marth. 2003. A genetic approach to mammalian glycan function. *Annual Review of Biochemistry* 72: 643–91.

Lynn, K. S., C. C. Chen, T. M. Lih, et al. 2015. MAGIC: An automated N-linked glycoprotein identification tool using a Y1-ion pattern matching algorithm and in silk MS2 approach. *Analytical Chemistry* 87: 2466–73.

Marino, K., J. Bones, J. J. Kattla, P. M. Rudd. 2010. A systematic approach to protein glycosylation analysis: A path through the maze. *Nature Chemical Biology* 6: 713–23.

Neelamegham, S., L. Gang, L. Y. Chi. 2015. GLYCOPAT: An open-source data analysis platform for individualized glycoproteome analysis. *Glycobiology* 25: 1282–3.

Ohtsubo, K., J. D. Marth. 2006. Glycosylation in cellular mechanisms of health and disease. *Cell* 126: 855–67.

Pap, A., E. Klement, E. Hunyadi-Gulyas, Z. Darula, K. F. Medzihradszky. 2018. Status report on the high-throughput characterization of complex intact O-glycopeptide mixtures. *Journal of the American Society for Mass Spectrometry* 29: 1210–20.

Park, G. W., J. Y. Kim, H. Hwang, et al. 2016. Integrated GlycoProteome analyzer (I-GPA) for automated identification and quantitation of site-specific N-glycosylation. *Scientific Reports* 6: 21175.

Park, G. W., J. W. Lee, H. K. Lee, J. H. Shin, J. Y. Kim, J. S. Yoo. 2020. Classification of mucin-type O-glycopeptides using higher-energy collisional dissociation in mass spectrometry. *Analytical Chemistry* 92: 9772–81.

Parker, B. L., M. Thaysen-Andersen, N. Solis, et al. 2013. Site-specific glycan-peptide analysis for determination of N-glycoproteome heterogeneity. *Journal of Proteome Research* 12: 5791–800.

Perez, S., A. Sarkar, A. Rivet, C. Breton, A. Imberty. 2015. Glyco3D: a portal for structural glycosciences. *Methods in Molecular Biology* 1273: 241–58.

Pinho, S. S., C. A. Reis. 2015. Glycosylation in cancer: Mechanisms and clinical implications. *Nature Reviews Cancer* 15: 540–55.

Pompach, P., K. B. Chandler, R. Lan, N. Edwards, R. Goldman. 2012. Semi-automated identification of N-glycopeptides by hydrophilic interaction chromatography, nano-reverse-phase LC-MS/MS, and glycan database search. *Journal of Proteome Research* 11: 1728–40.

Ranzinger, R., S. Herget, T. Wetter, C. W. von der Lieth. 2008. GlycomeDB-integration of open-access carbohydrate structure databases. *BMC Bioinformatics* 9: 384.

Savitski, M. M., T. Mathieson, I. Becher, M. Bantscheff. 2010. H-score, a mass accuracy driven rescoring approach for improved peptide identification in modification rich samples. *Journal of Proteome Research* 9: 5511–6.

Shakhsheer, B., M. Anderson, K. Khatib, et al. 2013. SugarBind database (SugarBindDB): A resource of pathogen lectins and corresponding glycan targets. *Journal of Molecular Recognition* 26: 426–31.

Soliman, C., E. Yuriev, P. A. Ramsland. 2017. Antibody recognition of aberrant glycosylation on the surface of cancer cells. *Current Opinion in Structural Biology* 44: 1–8.

Stadlmann, J., J. Taubenschmid, D. Wenzel, et al. 2017. Comparative glycoproteomics of stem cells identifies new players in ricin toxicity. *Nature* 549: 538–42.

Stadlmann, J., D. M. Hoi, J. Taubenschmid, K. Mechtler, J. M. Penninger. 2018. Analysis of PNGase F-resistant N-glycopeptides using SugarQb for proteome discoverer 2.1 reveals cryptic substrate specificities. *Proteomics* 18: 1700436.

Sun, S. S., Y. W. Hu, L. Jia, et al. 2018. Site-specific profiling of serum glycoproteins using N-linked glycan and glycosite analysis revealing atypical N-glycosylation sites on albumin and alpha-1B-glycoprotein. *Analytical Chemistry* 90: 6292–9.

Taus, T., T. Kocher, P. Pichler, et al. 2011. Universal and confident phosphorylation site localization using phosphoRS. *Journal of Proteome Research* 10: 5354–62.

Wang, L. H., D. Q. Li, Y. Fu, et al. 2007. pFind 2.0: A software package for peptide and protein identification via tandem mass spectrometry. *Rapid Communications in Mass Spectrometry* 21: 2985–91.

Wang, Y., Xu, F. F., Chen, Y., Tian, Z. X. 2020. A quantitative N-glycoproteomics study of cell-surface N-glycoprotein markers of MCF-7/ADR cancer stem cells. *Analytical and Bioanalytical Chemistry* 412: 2423–2432.

Waniwan, J. T., Y. J. Chen, R. Capangpangan, S. H. Weng, Y. J. Chen. 2018. Glycoproteomic alterations in drug-resistant non-small cell lung cancer cells revealed by lectin magnetic nanoprobe-based mass spectrometry. *Journal of Proteome Research* 17: 3761–73.

Woodin, C. L., D. Hua, M. Maxon, K. R. Rebecchi, E. P. Go, H. Desaire. 2012. GlycoPep grader: A web-based utility for assigning the composition of N-linked glycopeptides. *Analytical Chemistry* 84: 4821–9.

Xiao, K., Z. Tian. 2019. GPSeeker enables quantitative structural N-glycoproteomics for site- and structure-specific characterization of differentially expressed N-glycosylation in hepatocellular carcinoma. *Journal of Proteome Research* 18: 2885–95.

Xiao, K. J., Y. Wang, Y. Shen, Y. Y. Han, Z. X. Tian. 2018. Large-scale identification and visualization of N-glycans with primary structures using GlySeeker. *Rapid Communications in Mass Spectrometry* 32: 142–8.

Xiao, K. J., F. Yu, Z. X. Tian. 2017. Top-down protein identification using isotopic envelope fingerprinting. *Journal of Proteomics* 152: 41–7.

Xue, Y., J. J. Xie, P. Fang, et al. 2018. Study on behaviors and performances of universal N-glycopeptide enrichment methods. *Analyst* 143: 1870–80.

Xue, B. B., K. J. Xiao, Y. Wang, Z. X. Tian. 2020. Site- and structure-specific quantitative N-glycoproteomics study of differential N-glycosylation in MCF-7 cancer cells. *Journal of Proteomics* 212: 103594.

Yang, G. L., Y. W. Hu, S. S. Sun, et al. 2018. Comprehensive glycoproteomic analysis of Chinese hamster ovary cells. *Analytical Chemistry* 90: 14294–302.

Yang, J., W. Wang, Z. Chen, et al. 2020. A vaccine targeting the RBD of the S protein of SARS-CoV-2 induces protective immunity. *Nature* 586: 572–7.

Yin, H., M. R. An, P. K. So, M. Y. M. Wong, D. M. Lubman, Z. P. Yao. 2018. The analysis of alpha-1-antitrypsin glycosylation with direct LC-MS/MS. *Electrophoresis* 39: 2351–61.

Yu, Z. X., X. Y. Zhao, F. Tian, et al. 2017. Sequential fragment ion filtering and endoglycosidase-assisted identification of intact glycopeptides. *Analytical and Bioanalytical Chemistry* 409: 3077-87.

Zhang, H., P. Loriaux, J. Eng, et al. 2006. UniPep - a database for human N-linked glycosites: A resource for biomarker discovery. *Genome Biology* 7: R73.

Zhang, Y., Lin, T. H., Zhao, Y., Mao, Y. H., Tao, Y. R., Huang, Y., Wang, S. S., Hu, L. Q., Cheng, J. Q. & Yang, H. 2020. Characterization of N-linked intact glycopeptide signatures of plasma IgGs from patients with prostate carcinoma and benign prostatic hyperplasia for diagnosis pre-stratification. *Analyst* 145: 5353–62.

Zhang, Y., X. F. Xie, X. Y. Zhao, et al. 2018. Systems analysis of singly and multiply O-glycosylated peptides in the human serum glycoproteome via EThcD and HCD mass spectrometry. *Journal of Proteomics* 170: 14–27.

Zhao, S., I. Walsh, J. L. Abrahams, et al. 2018. GlycoStore: A database of retention properties for glycan analysis. *Bioinformatics* 34: 3231–2.

Zhu, Z. K., D. Hua, D. F. Clark, E. P. Go, H. Desaire. 2013. GlycoPep detector: A tool for assigning mass spectrometry data of N-linked glycopeptides on the basis of their electron transfer dissociation spectra. *Analytical Chemistry* 85: 5023–32.

Zhu, J. H., Z. W. Chen, J. Zhang, et al. 2019. Differential quantitative determination of site-specific intact N-glycopeptides in serum haptoglobin between hepatocellular carcinoma and cirrhosis using LC-EThcD-MS/MS. *Journal of Proteome Research* 18: 359–71.

CHAPTER 6

Clinical Applications

Caiyun Fang and Haojie Lu

CONTENTS

DOI: 10.1201/9781003185833-6

6.1 INTRODUCTION

Protein glycosylation plays a pivotal role in many important biological and pathological processes, such as molecular recognition, adhesion, signalling, etc. (de-Freitas-Junior et al. 2017; Jia et al. 2018; Pan et al. 2016; Rodrigues et al. 2018). Therefore, systematic study of protein glycosylation events has become indispensable in order to further understand various cellular processes and quest better solutions to biomedical problems. With the comprehensive advancements in qualitative and quantitative N-glycoproteomics and N-glycomics, which can provide a wealth of valuable information about expression level of glycoproteins and glycoforms, glycosylation sites, glycan structures and glycosite occupancy, state-of-the-art platforms have been widely applied in biological research, disease diagnosis, quality control of glycoprotein pharmaceuticals, and so on. This chapter focuses on the applications of mass spectrometry (MS)-based N-glycoproteomics and N-glycomics in clinical applications between 2015 and 2020, including glycoproteome and glycome profiling, potential biomarker and drug target discovery together with study of molecular mechanisms, as well as characterization of therapeutic glycoproteins (shown in Figure 6.1).

6.2 N-GLYCOPROTEOME AND N-GLYCOME PROFILING

As the most abundant protein post-translational modifications (PTMs), N-glycosylation plays vital roles in determining protein structure, function, and stability, and regulates nearly every physiological process. It is estimated that over half of human proteins are glycosylated, hence N-glycoproteome and glycome profiling is of great interest from a diagnostic and therapeutic point of view to further understand their functions and potential molecular mechanism. Proteins can be digested into glycopeptides using a proteolytic enzyme (e.g., trypsin), and then N-glycans can

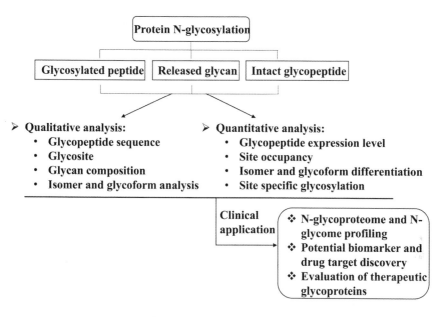

FIGURE 6.1 Overview of protein N-glycosylation analysis and its clinical applications.

be released from proteins by digestion with N-glycosidase F (PNGase F). Thus, protein glycosylation can be characterized via three different stages: the glycopeptides without glycans, the released glycans, or the intact glycopeptides. With the development of analytical methods, more and more N-glycosylated proteins and N-glycans have been identified. Here, we take commonly used clinical specimens, including body fluids and tissues, as an example to introduce the research progress.

6.2.1 Body Fluids

Body fluids (e.g., serum, saliva, and urine) are one of the most commonly used samples in glycoproteomics and glycomics because they are easily accessible and can often provide valuable information about the state of health of an individual or even serve as excellent noninvasive biomarkers for various diseases (Jóźwik and Kałużna-Czaplińska 2016).

6.2.1.1 Serum, Plasma, and Blood

For in-depth N-glycoproteome and N-glycome research, various selective enrichment and specific derivatization methods have been developed and applied in analysis of clinical samples including serum/plasma samples due to the inherent low stoichiometry and microheterogeneity of glycosylation, as well as low ionization efficiency in mass spectrometry. For example, Deng group developed a series of nanoparticles by optimizing their physical and chemical properties to improve enrichment performance for glycopeptides, such as zwitterionic hydrophilic magnetic nanoparticles (Wu et al. 2016a; Wu et al. 2016b), hydrophilic magnetic mesoporous silica materials (Sun et al. 2017), boronic acid functionalized magnetic microspheres (Wang et al. 2015; Xie et al. 2018), and so on. They could identify 424 glycopeptides assigned to 140 glycoproteins from 2 μL human serum (Sun et al. 2017). In Zhang group, different nanocomposites were synthesized not only to be directly used as enrichment materials for glycopeptides (Jiang et al. 2016a) but also to be constructed into monolithic columns (Jiang et al. 2016b; Liang et al. 2015), contributing to online analysis of glycopeptides. The 262 unique N-glycosylated peptides corresponding to 124 N-glycoproteins could be identified from 1 μL human serum using their amide functionalized hydrophilic monolithic capillary column (Jiang et al. 2016b). At the same time, various analytical strategies have also been used and achieved good results. For instance, Cao et al. (2016) developed the zwitterionically functionalized Poly (amidoamine) dendrimer (ZICF-PAMAM) combined with filter-aided sample preparation (FASP)-mode enrichment strategy, which could efficiently enrich glycopeptides from complex biological samples even for merely 0.1 μL human serum samples; Jie et al. (2020) used hydrophilic interaction liquid chromatography (HILIC)-PNGase F-HILIC strategy to identify 722 N-glycopeptides within 202 unique glycoproteins from 1 μL human plasma digest. To facilitate the separation and/or MS detection of glycome, Zhao et al. (2015) synthesized hydrazino-s-triazine-based labelling reagents to enhance MS signal and identified 62 N-glycans released from human serum glycoproteins. Wang et al. (2019) detected 55 N-glycan compositions from 0.1 μL human serum sample by developing a glycan reductive amino acid coded affinity tagging method, including some N-glycans with low abundance such as tetra-antennary H7N6S2, H7N6F1S3, and H7N6F1S4. To simplify analytical procedure and improve analytical throughput, Sajid et al. (2020) proposed a tip-based strategy and detected a total of 59 N-glycans derived from serum glycoproteins. Bladergroen et al. (2015) presented a fully automated protocol based on a 96-well filter plate with GH Polypro membrane for the high-throughput analysis of plasma N-glycans, in which 384 samples could be prepared automatically within a running time of 5.5 hours, followed by matrix assisted laser desorption ionization-time of flight mass spectrometry (MALDI-TOF MS) analysis within another 1.5 hours. This kind of robotic sample preparation system could bring high reproducibility and high throughput compared to that of manual processing, being suitable for the analysis of clinical samples.

Glycoproteome and glycome in serum are complicated. Song et al. (2015) monitored the N-glycan variation in serum among nine individuals from the most abundant to the least. These glycans represented more than four orders of magnitude and corresponded to more than 170 structures. They found that the most abundant N-glycan was not from immunoglobulin G (IgG), the most abundant glycoprotein, but originated from several glycoproteins, and the abundances were distributed throughout many glycan structures. Sun et al. (2018) observed that half of N-glycosites in their serum N-glycoproteome dataset were modified by at least two glycans, and a majority of them were sialylated. Specifically, bi-antennary N-glycans were found in 3/4 of glycosites, and tri-antennary sialylated N-glycans were found in 1/3 of glycosites.

Besides, serum/plasma samples can be further subdivided and analyzed, such as human peripheral blood mononuclear cells and lymphocytes (Chen et al. 2018b) and extracellular vesicles (Aguilar et al. 2020). In addition, dried blood spot samples are also of interest due to its ease of collection, transportation, and storage. Choi et al. (2017) directly identified a total of 41 site-specific N-glycopeptides from 16 glycoproteins (from IgG1 (10 mg/mL) down to complement component C7 (50 µg/mL)) in the dried blood spot samples. They found that there were 32 N-glycopeptides from 14 glycoproteins that could be consistently quantified over 180 days even when stored at room temperature, indicating that the dried blood spot samples probably have a good application prospect in the future.

6.2.1.2 Urine

Zhang et al. identified an average of 1465 glycopeptides from 839 glycoproteins and 1553 glycopeptides from 884 glycoproteins from female and male urine samples in a single MS analysis after a hydrophilic material 4-mercaptobenzene boronic acid functionalized and Au-doped straticulate C3N4 enrichment (Zhang et al. 2019b), and captured 1809 human urine N-glycopeptides corresponding to 876 N-glycoproteins from two male and two female urine using zwitterionic hydrophilic L-cysteine derivatized straticulate-C3N4 composites (Zhang et al. 2020a).

Exosomes are released by almost all cell types containing proteins and nucleic acids and increasingly being recognized as important vehicles for intercellular communication (An et al. 2015a; Cocucci and Meldolesi 2015), and have attracted increasing attention for biomarker discovery and disease treatment in recent years (Bellin et al. 2019). Zou et al. (2017) detected high-mannose and complex N-glycans as well as pauci-mannosidic structures with high degree of fucosylation and sialylation in urinary exosomes collected from healthy male individuals. Chen et al. (2020b) identified intact glycopeptides in normal urine samples by combining tip-based hydrophilic and hydrophobic peptide extraction chemistries and a 96-well plate-based liquid handling platform, in which they observed the glycopeptides carrying sialylated glycans, oligomannose glycans, and fucosylated glycans as well as other complex glycans. Zhang et al. (2020c) identified 1250 N-glycopeptides from human urine exosome using their magnetic hydrophilic material MoS_2-Fe_3O_4-Au/NWs-GSH to enrich N-glycopeptides and found that the identified glycoproteins were most correlated with cell adhesion, cell-matrix adhesion, and platelet degranulation in biological process, while mostly associated with virus receptor activity, receptor activity, and cell adhesion molecule binding in molecular function.

6.2.1.3 Other Human Body Fluid Samples

N-glycoproteome and N-glycome in other human body fluid samples, such as saliva, amniotic fluid, cerebrospinal fluid, milk, and so on, have also been well characterized. For instance, saliva collection is noninvasive and relatively easy, making it an ideal sample for scientific research and disease diagnosis, and there is a relatively high proportion of glycosylated proteins in the salivary proteome. Hence, profiling of glycoproteome and glycome in saliva increasingly attracts researchers' interests (Caragata et al. 2016; Li et al. 2018; Zheng et al. 2020). Shi et al. (2017) identified a total of 126 N-linked glycopeptides corresponding to 97 glycoproteins in human amniotic fluid. Goyallon et al. (2015) identified 124 N-glycopeptides representing 55 N-glycosites from 36 glycoproteins in a pooled human cerebrospinal fluid sample by combining N-glycomics and N-glycoproteomics.

Besides qualitative characterization, quantification of glycoproteins and glycans via MS-based methods has also been reported. For example, Huang et al. (2017a) quantified seven milk proteins (α-lactalbumin, lactoferrin, secretory IgA, IgG, IgM, α1-antitrypsin, and lysozyme) using their unique peptides, as well as their site-specific N-glycosylation relative to the protein abundance, by a label-free multiple reaction monitoring (MRM) method.

6.2.2 Characterization of Specific Glycoproteins

6.2.2.1 Immunoglobulins (Igs)

Immunoglobulins are key components in the humoral immune system and known as glycoproteins. There are five types of immunoglobulins (IgG, IgA, IgM, IgD, and IgE) in our blood. Igs bind to antigens through their fragment antigen-binding (Fab) region, comprising two heavy chains and two light chains, interact with cell surface receptors through the fragment crystallized (Fc) region. Glycosylation on the Fab and Fc regions plays a key role in the regulation of immune reactions and modulates a diversity of immunoglobulin properties including protein conformation and stability, serum half-life, as well as binding affinities to antigens and receptors. Hong et al. (2015) used MRM technique to simultaneously quantify IgG, IgA, IgM, and their site-specific glycans in the serum of 13 healthy individuals and a pooled commercial serum sample (Sigma-Aldrich), in which a total of 64 glycopeptides and 15 peptides were monitored. They found that there were a large biological variation in glycan expression even within the general healthy individuals (CV>50% when using absolute glycopeptide abundance). However, there was a similar degree of glycosylation across samples and the biological variation decreased (CV<20%) if normalized to the protein content. In addition, because monoclonal antibodies (mAbs) are an effective therapeutic tool for a variety of diseases such as cancer and inflammatory diseases (Beck et al. 2012), the antibody drugs have constituted the fastest growing class of biotherapeutics (Walsh 2014). Due to their relevance as biotherapeutics, extensive characterization of glycosylation on Igs can not only increase molecular and structural understanding of IgG, but also drive the development of therapeutics and individualized treatment (Dalziel et al. 2014).

6.2.2.1.1 IgG

IgG consists of IgG1, IgG2, IgG3, and IgG4, and these four subclasses are highly conserved but have unique differences that result in subclass-specific effector functions. Bas et al. (2019) demonstrated that sialylation of the asparagine 297 (N297) sugar moiety could enhance human IgG serum persistence. Shaha et al. (2017) revealed some structural perturbations based on sequence differences, although the overall structure of IgG3 Fc was similar to that of other subclasses. Glycosylation on IgG has been often analyzed. Ma et al. (2016) detected 33 glycopeptides from the digest of human IgG. Huang et al. (2016) analyzed IgG isomeric N-glycopeptides in human serum that had the same peptide backbone but isomeric glycans, such as sialylated N-glycan isomers differing in $\alpha2,3$ and $\alpha2,6$ linkages using HILIC separation. Dong et al. (2016) found a total of 247 glycopeptide ions and 60 glycans of different masses in the tryptic digestion of an IgG1 reference product using 1D and 2DLC-MS/MS. Stavenhagen et al. (2015) investigated the glycosylation of human IgG3 using collision-induced dissociation with a combination of lower- and enhanced-energy, which could acquire information about both the glycan and the peptide moiety in one run, and found that IgG3 also possessed a N-linked site in the CH3 domain at Asn392 in addition to the well-known Fc N-glycosylation site. Furthermore, absolute quantitation of IgG1 Fc-glycosylation is crucial for the clinical practice of glycobiomarkers and quality control of biopharmaceuticals. Cao et al. (2020) developed an absolute quantitation strategy and determined 11 high abundant IgG1 Fc-glycopeptides with definite peptide sequences and glycoforms in pooled human sera from 70 healthy subjects. They found a wide range of IgG1 Fc-glycopeptide concentrations from 0.60 to 17.61 nmol/mL.

It should be mentioned that different IgG glycoforms exhibited a different susceptibility to tryptic cleavage beyond the simple difference between glycosylated and nonglycosylated proteoforms, although trypsin is by far the most commonly used protease in glycoproteomic studies. Falck et al. (2015) observed a strong preferential digestion of high mannose, hybrid, $\alpha2,3$-sialylated and bisected glycoforms over the most abundant neutral and fucosylated glycoforms. And this bias was dependent on the intact higher order structure of the antibodies to a large extent. In addition, some unspecific proteases such as pronase may also be a good choice, because it can generate glycopeptides with a small peptide portion so as to facilitate the site and glycan identification. During the process of production, storage, and transportation, exposure to ambient light is inevitable, but it can cause protein physical and chemical degradation. Kang et al. (2019) demonstrated that the UV exposure of wild-type IgG4 Fc could lead to the side-chain cleavage of specific tyrosine residues, which might be related to light-induced inactivation, aggregation, fragmentation, or immunogenicity of protein therapeutics.

6.2.2.1.2 IgA

IgA can be further classified into IgA1 and IgA2. Huang et al. (2015) analyzed the site-specific glycosylation of secretory IgA isolated from human colostrum (n = 3) and found that the majority of the glycans were bi-antennary structures with one or more acidic Neu5Ac residue, but a large fraction contained truncated complex structures with terminal GlcNAc. To gain an insight into oral cavity-specific antibody glycosylation, Plomp et al. (2018) characterized the glycosylation of salivary (secretory) IgA, including the IgA joining chain (JC) and secretory component (SC). IgG and IgA were affinity-purified from the human plasma and saliva samples collected from 19 healthy volunteers within a 2-hour time window, followed by tryptic digestion and nanoLC-ESI-Q-TOF MS analysis to compare their glycosylation. They found that saliva-derived IgG exhibited a slightly lower galactosylation and sialylation than plasma-derived IgG had. Additionally, glycosylation of IgA1, IgA2, and the JC showed substantial differences between plasma and saliva samples. For example, salivary proteins exhibited a higher bisection, and lower galactosylation and sialylation as compared to plasma-derived IgA and JC. All seven N-glycosylation sites on the SC of secretory IgA in saliva carried highly fucosylated and fully galactosylated di-antennary N-glycans.

6.2.2.1.3 IgM

Pabst et al. (2015) found that human serum IgM showed site-specific glycosylation: more than 95% of glycoforms on site Asn171 were complex type structures; glycosite Asn563 carried solely oligomannosidic structures; Asn402, 332, and 171 had small amounts of hybrid-type glycans; Asn395 and the J chain only presented complex-type glycans with one or two sialic acid residues and a bisecting GlcNAc.

6.2.2.2 *Other Glycoproteins*

Reiding et al. (2019a) structurally characterized myeloperoxidase from neutrophils of healthy human donors by combining bottom-up glyco-proteomics and native MS approaches, which contained five putative N-glycosylation sites (ten in the mature dimer). They found that Asn355 and Asn391 mostly harboured high-mannose-type glycans, whereas Asn483 was the dominant site with complex-type species. Solakyildirim et al. (2018) investigated N-glycan structures from human placental alkaline phosphatase using HILIC-Orbitrap MS and identified 16 structures including ten sialylated N-glycans. Velkova et al. (2017) investigated N-glycans of the structural subunit β-HlH of haemocyanin isolated from Helix lucorum. In total, 32 different glycans were enzymatically liberated and characterized by tandem MS using a Q-Trap mass spectrometer.

6.3 POTENTIAL BIOMARKER AND DRUG TARGET DISCOVERY

Protein glycosylation plays crucial roles in broad biological processes and activities (such as immune-response, signal transduction, cell differentiation, gene expression, protein degradation, and so on) (Varki 2017), while aberrant protein glycosylation has close association with the occurrence and development of a variety of diseases (e.g., various cancers, neurological disorders, etc.) (Li et al. 2019; Vajaria and Patel 2017; Veillon et al. 2018). Therefore, glycoproteomic and glycomic studies are of great interest to analyze the abnormal changes of N-glycopeptides and glycans in clinical samples for not only the development of disease biomarkers and drug target discovery, but also providing information regarding underlying mechanisms of diseases. For example, the majority of cancer biomarkers approved by the Food and Drug Administration (FDA) are glycoproteins, such as α-fetoprotein (AFP), prostate-specific antigen (PSA), carcinoembryonic antigen, and carbohydrate antigen (CA) 15-3, etc. (Kailemia et al. 2017; Kirwan et al. 2015; Munkley and Elliott 2016; Silsirivanit 2019). Different clinical specimens are used due to their advantages and disadvantages (shown in Figure 6.2).

6.3.1 Body Fluids

6.3.1.1 Serum, Plasma, and Blood

Human plasma/serum is most commonly used as clinical samples, which can reflect the immediate physiology and possesses significant potential for disease diagnosis as well as therapeutic monitoring. Therefore, global profiling of human serum and plasma N-glycoproteome and N-glycome has become a noninvasive method for disease biomarker discovery.

6.3.1.1.1 Cancers

More and more studies have demonstrated that abnormal glycosylation is closely related to various cancers. Table 6.1 lists some examples.

Ruhaak et al. (2016a) analyzed serum glycan profiles to distinguish non-small cell lung cancer (NSCLC) cases from controls, in which a

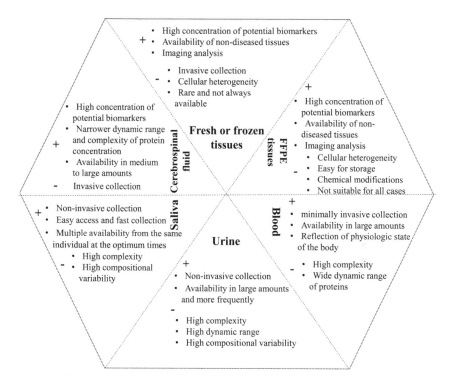

FIGURE 6.2   Clinical samples used for biomarker and drug target discovery, along with their advantages and limitations.

discovery set included 100 NSCLC cases and 199 healthy controls, and a second test set consisted of 108 cases and 216 controls. They found the serum glycans were related with risk assessment for NSCLC. Twelve glycans exhibited >0.6 area-under-the-curve (AUC) in the discovery set, showing significant discriminatory power between cases and controls. When combined with pro-surfactant protein B (pro-SFTPB), four glycans exhibited significant incremental value with AUCs of 0.73, 0.72, 0.72, and 0.72 in the test set. Zhang et al. (2019c) studied the human serum N-glycome of 91 multiple myeloma patients and 51 controls, sequential samples from patients (n = 7) obtained at different time points during disease development as well as 16 paired blood serum and bone marrow plasma samples. They found, compared to controls, a decrease in both α2,3- and α2,6-sialylation, galactosylation and an increase in fucosylation within complex-type N-glycans in multiple myeloma patients. Furthermore, difucosylation of di-antennary glycans

TABLE 6.1   List of Some Clinical Applications of N-Glycoproteomics and N-Glycomics in Cancer-Associated Biomarker and Drug Target Discovery by Using Serum, Plasma, or Blood as Samples

Cancer Types	References	Cancer Types	References
Lung cancer	Ruhaak et al. 2016a	Pancreatic cancer	Schauer 2009; Li and Ding 2019; Liu et al. 2018a; Li et al. 2016; Tan et al. 2015
Multiple myeloma	Zhang et al. 2019c	Hepatocellular carcinoma	Zhao et al. 2020; Yang et al. 2019; Yin et al. 2015; Tanabe et al. 2016; Liu et al. 2017a; Ma et al. 2018; Liu et al. 2017b; Gebrehiwot et al. 2018
Esophageal adenocarcinoma	Shah et al. 2015		
Oral squamous cell Carcinoma	Guu et al. 2017	Ovarian cancer	Kim et al. 2017c; Dĕdová et al. 2019; Biskup et al. 2017
Gastric cancer	Qin et al. 2017b; Wang et al. 2020b	Colorectal cancer	Peng et al. 2019; de Vroome et al. 2018
Papillary thyroid carcinoma	Zhang et al. 2020b	Cholangiocarcinoma	Talabnin et al. 2018
Breast cancer	Hu and Borges 2017	Germ-cell tumors	Narita et al. 2017

The columns represent the cancer type and the reference number in the review.

decreased with development of the disease in individual patients. There was strong correlation between protein N-glycosylation of blood and that of bone marrow. Guu et al. (2017) investigated serum N-glycomes and anti-carbohydrate antibodies from normal populations and oral squamous cell carcinoma (OSCC) patients. They found that the diagnostic accuracy of seven N-glycans (e.g., di-sialylated bi-antennary glycan, di-sialylated tetra-antennary glycan, etc.), which were decreased or increased in serum of OSCC, were greater than 75%. The relative abundances of total tri-antennary and tetra-antennary glycans with different degrees of fucosylation and sialylation were also increased in serum of OSCC patients. Furthermore, in an independent validation group of 48 OSCC patients, most of the high-molecular weight serum N-glycans exhibited significantly high sensitivity and specificity. The levels of two

IgM antibodies were elevated, but the levels of nine IgG antibodies were decreased in patient serum. Qin et al. (2017b) identified 81 N-glycans (peaks) in 203 serum samples (163 gastric cancer patients and 40 controls) using matrix-assisted laser desorption/ionization mass spectrometry (MALDI MS). They observed that hybrid and multi-branched (tri- or tetra-antennary) N-glycans were increased; whereas four other types of N-glycans (monoantennary, galactose, bisecting, and core fucose) were decreased in gastric cancer. Core fucose exhibited an excellent diagnostic performance for the early detection with AUC 0.923, 95% confidence interval (CI) 0.8485 to 0.9967 in the training set, and its diagnostic potential was AUC 0.854, 95% CI 0.7592 to 0.9483 in an independent cohort. Wang et al. (2020b) found that bisecting GlcNAc and tri-antennary glycan compositions in serum were significantly changed between gastric cancer cases (n = 50) and healthy control. Shah et al. (2015) carried out a serum glycoprotein biomarker discovery and candidate qualification for Esophageal Adenocarcinoma (EAC). There were 29 samples (healthy, n = 9; Barrett's oesophagus (BE), n = 10; and EAC, n = 10) in biomarker discovery phase and 79 samples (healthy, n = 20; BE, n = 20; EAC, n = 20; and population control, n = 19) for biomarker qualification study, in which EAC developed from metaplastic condition BE. Of the 246 glycoforms measured in the qualification stage, 40 glycoforms were considered to qualify as candidate serum markers. The top candidate for distinguishing healthy from BE patients' group was *Narcissus pseudonarcissus* lectin-reactive Apolipoprotein B-100; the candidate for distinguishing BE from EAC was *Aleuria aurantia* lectin-reactive complement component C9; and the *Phaseolus vulgaris* Erythroagglutinin (PHA-E)-reactive gelsolin was for distinguishing healthy from EAC. A panel of 8 glycoforms showed an improved AUC (0.94) to discriminate EAC from BE. Talabnin et al. (2018) found that the expression of N-glycans mannose 6-N-acetyl-glucosamine 2, mannose 9-N-acetyl-glucosamine 2, and NeuAc3H3N3M3N2F in plasma samples differed significantly between the cholangiocarcinoma cases and healthy controls. Hu and Borges (2017) observed three glycan nodes (2,4-linked mannose (2,4-Man), 2,6-linked mannose (2,6-Man), and 4,6-linked GlcNAc (4,6-GlcNAc) with significantly different distributions between the blood plasma samples from stage III-IV breast cancer patients (n = 20) and age-matched controls (n = 20) (ROC (receiver operating characteristic) >0.75; p<0.01), which might potentially be useful as markers for stage III-IV breast cancer. Narita et al. (2017) carried out a comprehensive serum N-glycan structural analysis from 54 untreated germ-cell tumours (GCT) patients and 103 age-adjusted healthy volunteers using glycoblotting methods and MS. Five candidate N-glycans were identified to be significantly associated with GCT patients with an AUC value of 0.87. Diagnostically, the combination of GCT-related N-glycans detected 10 of 12 (83%) patients with negative conventional tumour markers. Prognostically, the predictive value of the prognostic N-glycan score was with an AUC value of 0.89. In addition, they thought that the GCT-related N-glycans were not strongly associated with immunoglobulins. Zhang et al. (2020b) explored human papillary thyroid carcinoma (PTC)-related proteins and glycosylations with reference to healthy controls (n = 5). They established a human plasma N-glycoproteomic database, containing 369 N-glycoproteins, 862 glycosites, 171 glycan compositions, and 1644 unique intact N-glycopeptides. They found that most of the proteins had multiple glycosites with various glycans. For example, the glycosite Asn65 and Asn98 in apolipoprotein D from healthy control plasma were modified by 5 and 27 different glycans, respectively. And nearly 20% of intact glycopeptides were different between healthy and PTC plasma.

For hepatocellular carcinoma (HCC), Zhao et al. (2020) found that the abundance of glycans containing galactosylation, core fucosylation, and sialic acid changed significantly in HCC patient serum. Yang et al. (2019) quantified a total of 30 N-glycan compositions in the HCC and normal serum, and found that 15 N-glycan compositions with bisecting GlcNAc, sialic acid, and core fucosylation exhibited significant differences. Structure-specific glycan profiling may provide more potential biomarkers with higher specificity. Hence, Yin et al. (2015) identified 1300 core fucosylation peptides from 613 serum proteins in early stage HCC with different aetiologies. They found that 20 core fucosylation peptides were differentially expressed in alcohol-related HCC samples compared with alcohol-related cirrhosis samples, and 26 core fucosylation peptides changed in hepatitis C virus (HCV)-related HCC samples compared with HCV-related cirrhosis samples. Three core fucosylation peptides from fibronectin (sites 528, 542, and 1007) up-regulated in alcohol-related HCC samples compared with cirrhosis samples. All three sites exhibited very high AUC values. Especially, the AUC value at site 1007 was 0.89 with a specificity 85.7% at a sensitivity of 92.9%, while the AUC value could reach to 0.92 with a specificity of 92.9% at a sensitivity of 100% when combined with the AFP index, which was significantly improved compared to that with AFP alone. Tanabe et al. (2016) analyzed sera from 42 HCC patients and 80 controls (composed of 27 chronic hepatitis B patients, 26 chronic hepatitis C patients, and 27 healthy volunteers) and revealed that α1-acid glycoprotein with multifucosylated tetra-antennary N-glycans was significantly elevated in HCC patients, whereas the single fucosylated derivative was not. Liu et al. (2017a) enriched and quantified glycoproteins with the sialylation α2,3-Gal structure in serum samples from HBV-related HCC patients and healthy controls using *Maackia amurensis* lectin (MAL) affinity chromatography and isobaric tag for relative and absolute quantitation (iTRAQ) labelling technique. Seventeen differential MAL-associated glycoproteins were identified. Especially, Galectin-3-binding protein was verified as one of the potential serological biomarkers for diagnosing HBV-related HCC, and its diagnostic accuracy was further enhanced when combined with serum AFP. Ma et al. (2018) quantified the difference in core fucosylated N-glycopeptides of serum proteins among three comparison groups (15 healthy control, 15 fibrosis, and 15 cirrhosis patients) using a LC-MS-MRM method. They found core fucosylation of five glycopeptides (N630 of serotransferrin, N107 of alpha-1-antitrypsin, N253 of plasma protease C1 inhibitor, N397 of ceruloplasmin, and N86 of vitronectin) was increased at the stage of liver fibrosis, and six glycopeptides (N138 and N762 of ceruloplasmin, N354 of clusterin, N187 of hemopexin, N71 of immunoglobulin J chain, and N127 of lumican) were increased at the stage of cirrhosis. Liu et al. (2017b) quantified the changes in N-glycosite occupancy for HCC metastasis serum. They identified 11 glycoproteins with significantly changed N-glycosite occupancy, which were associated with cell migration, invasion, and adhesion through p38 mitogen-activated protein kinase signalling pathway and nuclear factor kappa B signalling pathway. Gebrehiwot et al. (2018) analyzed the serum N-glycomes of 54 healthy subjects of various ethnicities (the United States (US), South Indian, Japanese, and Ethiopian) and 11 Japanese HCC patients using glycoblotting-assisted MALDI-TOF-MS-based quantitative analysis. Among the total of 51 N-glycans quantified reproducibly, 33 glycoforms were detected in all ethnicities. The 13 N-glycans were detected weakly but exclusively in the Ethiopians and 5 glycans in all the other ethnic groups. Their results demonstrated an ethnic-specific expression pattern of N-glycome with marked alterations in their serum abundance, emphasizing that ethnicity matching should be considered for accurate glyco-biomarker identification.

About ovarian cancer (OVC), Kim et al. (2017c) compared human serum N-glycans between healthy controls and OVC patients, including a training set (40 healthy controls and 40 OVC cases) and a blind test set (23 controls and 37 patients). They found that sensitivity of CA-125 (the most widely used OVC marker) was 74% in the training set and 78% in the test set, respectively. By contrast, a multibiomarker panel composed of 15 MALDI-TOF-MS peaks resulted in AUC of 0.89, 80~90% sensitivity, 70~83% specificity in the training set, and 81~84% sensitivity, 83% specificity in the test set, indicating potential usefulness for the screening of OVC. Dědová et al. (2019) observed a statistically significant decrease for high-mannose, hybrid-type, complex-type asialylated, bi-, tri-, and tetra-antennary sialylated structures was between late stage patients and controls or early stage OVC patients. Biskup et al. (2017) analyzed the N-glycome profiles of ascitic fluid from primary serous epithelial ovarian cancer (EOC) patients and compared them with the serum N-glycomes of 18 EOC patients and 20 age-matched healthy controls. Seven N-glycans (one bi-antennary structure, two sialylated tri-antennary difucosylated structures, one tetra-antennary structure, and three tetra-antennary fucosylated structures) showed significant differential expression among all three sample cohorts. Ascites N-glycome showed increased antennarity, branching, sialylation, and Lewis X motifs compared to healthy serum.

Sialylated N-glycans play pivotal roles in several important biological and pathological processes (Li and Ding 2019; Schauer 2009), while sialyl-linkage isomers, mostly α2,3- and α2,6-linked, act differently during the cellular events and several diseases. Peng et al. (2019) found an increase in total α2,6-sialic acid level in serum N-glycans of colorectal cancer (CRC) patients (ten healthy individuals and ten CRC patients). de Vroome et al. (2018) investigated CRC associated changes of the serum N-glycome in a matched case-control study (124 cases vs 124 controls) and then validated their results in an independent sample cohort (61 cases vs 61 controls) and in post-operative samples from cured CRC cases. They found increased size (antennae) and sialylation of the N-glycans in the CRC patient sera as compared to mainly di-antennary N-glycans in sera from controls. And glycan alterations have strong associations with cancer stage and survival. Liu et al. (2018a) compared relative abundances of over 280 N-glycan isomers in sera between pancreatic cancer (PC) cases (n = 32) and healthy controls (n = 32) using nanoLC-ESI-MS. Twenty five specific-isomeric biomarkers were found to be significantly different (p<0.05), among which the majority of significantly altered isomers were sialylated. Li et al. (2016) found that the tri-antennary glycans between $\alpha(2,3,6,6)$ and $\alpha(2,6,6,6)$-linked sialic acid structures had the significant difference between PC and control serum samples and the ratio of the tri-antennary glycans with $\alpha(2,6,6,6)$ linked sialic acids elevated in the pancreatic samples. Fucosylation is another important type of glycosylation involved in cancers. Tan et al. (2015) identified 630 sites containing core fucosylated structures in 322 proteins in the depleted serum samples using the Orbitrap Elite MS, and observed 8 potential core fucosylated glycopeptide cancer markers that differed between PC and healthy controls or chronic pancreatitis.

To examine blood protein changes in early-stage localized cancers, Sajic et al. (2018) analyzed 284 blood samples from patients with 5 types of localized-stage carcinomas (colorectal, pancreatic, lung, prostate, and ovarian carcinomas) and healthy controls using a proteomic workflow combining N-glycosite enrichment and SWATH MS. They discovered that proteins related to blood platelets were common to several cancers (e.g., THBS1), whereas others were highly cancer-type specific.

Due to their potentially important cellular functions in disease onset and progression, extracellular vesicles (EVs) are increasingly recognized as valuable sources for biomarker discovery and disease diagnosis (Bae et al. 2018; Emmanouilidi et al. 2019; Hurwitz and Meckes 2019; Wu et al. 2019). For instance, Chen et al. (2018a) identified 1453 unique glycopeptides representing 556 glycoproteins in EVs, among which 20 were verified significantly higher in individual breast cancer patients. Bai et al. (2018) identified 329 N-glycosylation sites corresponding to 180 N-glycoproteins from plasma exosomes of glioma patients and healthy subjects, among which 26 N-glycoproteins were significantly changed. Melo et al. (2015) identified a cell surface proteoglycan, glypican-1 (GPC1), specifically enriched on cancer-cell-derived exosomes, because GPC1[+] circulating exosomes (crExos) were detected in the serum of patients with pancreatic cancer with absolute specificity and sensitivity, distinguishing healthy subjects and patients with a benign pancreatic disease from patients with early- and late-stage pancreatic cancer, and levels of GPC1[+] crExos correlated with tumour burden and the survival of pre- and post-surgical patients.

6.3.1.1.2 Inflammatory Diseases

DeCoux et al. (2015) identified a total of 501 N-linked plasma glycopeptides corresponding to 234 proteins from the Lactate Assessment in the treatment of early sepsis cohort by solid-phase extraction coupled with MS. Of these, 66 glycopeptides were unique to the survivor group corresponding to 54 proteins, 60 were unique to the nonsurvivor group corresponding to 43 proteins, and 375 were common responses between groups corresponding to 137 proteins. Clerc et al. (2018) compared the plasma N-glycosylation profiles between patients with inflammatory bowel diseases (n = 2635) and healthy individuals (n = 996). They found that plasma samples from patients had a higher abundance of large-size glycans compared with controls, a decreased relative abundance of hybrid and high-mannose structures, lower fucosylation, lower galactosylation, and higher sialylation (α2,3- and α2,6-linked). Ząbczyńska et al. (2020) found that both the monosialylated tri-antennary glycan (A3G3S1; m/z 1209.40) and disialylated di-antennary structure with antennary Fuc (FA2G2S2; m/z 1245.42) showed a statistically significant increase in IgG-depleted serum samples from Hashimoto's thyroiditis patients compared with those from healthy donors. To identify protein N-glycosylation properties associated with rheumatoid arthritis (RA) disease activity during and after pregnancy, Reiding et al. (2018) studied the N-glycosylation of sera from 253 RA patients and 32 control pregnancies at 7 timepoints before, during, and after pregnancy by MALDI-TOF MS, in which they reported the disease- and pregnancy- associated changes of 78 N-glycan species and 91 glycosylation traits derived thereof. They found a decrease in bisection and an increase in galactosylation in di-antennary glycans, as well as an increase in tri- and tetra-antennary species and α2,3-linked sialylation. To carry out association analysis with markers of metabolic health and inflammation, Reiding et al. (2017) studied the total plasma N-glycome of 2144 healthy middle-aged individuals from the Leiden Longevity Study using matrix-assisted laser desorption/ionization-Fourier transform ion cyclotron resonance-MS (MALDI-FTICR-MS). They performed the relative quantification of 61 glycan compositions, ranging from $Hex_4HexNAc_2$ to $Hex_7HexNAc_6dHex_1Neu_5Ac_4$, and 39 glycosylation traits derived thereof. They found that the bisection, galactosylation, and sialylation of di-antennary species, the sialylation of tetra-antennary species, and the size of high-mannose species were associated with inflammation and metabolic health. Their results revealed higher levels of fucosylation in di-, tri-, and tetra-antennary glycan traits in males vs females, but the presumably IgG-related trait A2FS0B was found to be lower in men than in women. Pregnancy requires partial suppression of the immune system to ensure maternal-foetal tolerance. Protein glycosylation, especially terminal sialic acid linkages, are of prime importance in regulating the pro- and anti-inflammatory immune responses. Jansen et al. (2016) analyzed the serum N-glycome of a cohort of 29 healthy women at 6 timepoints during and after pregnancy. A total of 77 N-glycans were followed over time. They observed an increase during pregnancy and decrease after delivery for both $\alpha 2,3$- and $\alpha 2,6$-linked sialylation, as well as a difference in the recovery speed after delivery for $\alpha 2,3$- and $\alpha 2,6$-linked sialylation of tri-antennary glycans.

6.3.1.1.3 Other Diseases

Dong et al. (2018) compared the serum glycome of patients with idiopathic REM sleep behaviour disorder (iRBD) and healthy controls. Sixteen significantly altered N-glycans structures were identified ($p < 0.05$), among which N-glycans with the composition of $HexNAc_4Hex_5Fuc_1$, $HexNAc_5Hex_5$, and $HexNAc_4Hex_5Fuc_1NeuAc_1$ exhibited the most substantial difference between controls and RBD patients ($p < 0.01$). Among seven significantly different N-glycan isomers ($p < 0.05$), $HexNAc_4Hex_5Fuc_1NeuAc_1$ (4511-2) and $HexNAc_4Hex_5Fuc_1NeuAc_2$ (4512-2) showed the most substantial difference ($p < 0.001$) and were higher in the idiopathic RBD cohort. Qin et al. (2017c) compared serum glycopatterns and the MAL-II binding glycoproteins (MBGs) in 65 children with Autistic spectrum disorder (ASD) and 65 age-matched typically developing (TD) children. Compared with that from TD children, the expression of sialylation $\alpha 2,3$-Gal/GalNAc was significantly increased in pooled and individual serum samples from ASD. A total of 194 and 217 MBGs were identified from TD and ASD sera, respectively, of which 74 proteins were specially identified or up-regulated in ASD. Moreover, increase of APOD $\alpha 2,3$ sialoglycosylation could sensitively and specifically distinguish ASD samples from TD samples with AUC 0.88. Borelli et al. (2015) analyzed the plasma N-glycome of 76 Down Syndrome (DS) persons, 37 siblings, and 42 mothers of DS persons. They found an overall decrease of galactosylation and $\alpha 2,3$ sialylation in DS, companied by increase of the level of fucosylated N-glycans and monogalactosylated di-antennary N-glycans. The GlycoAgeTest and the ratio of two core-fucosylated, monogalactosylated di-antennary isomers (galactose positioned on $\alpha 1,6$ arm vs $\alpha 1,3$ arm) were the strongest DS discriminators. Hypogalactosylation was a characteristic for DS and the aging population. A decrease in $\alpha 2,3$-sialylated species was found both in DS and aging of controls. The $\alpha 2,6$-sialylated tri- and tetra-galactosylated N-glycans were lowered in DS but increased with age in the same persons, but they were not affected by aging in control group. Noro et al. (2017) analyzed N-glycan levels in the postoperative sera of 197 living donor kidney transplant recipients including 16 recipients who had antibody-mediated rejection (ABMR) with or without T-cell-mediated rejection (TCMR), 40 recipients with TCMR, 141 recipients without adverse events, and 135 healthy controls. The N-glycan score discriminated ABMR with 81.25% sensitivity, 87.85% specificity, and an AUC of 0.892. Recipients with N-glycan-positive scores >0.8770 had significantly shorter ABMR survival than that of recipients with N-glycan-negative scores. Saldova et al. (2015) characterized the glycosylation changes in both total serum glycoproteins and isolated serum IgG from ten previously reported MAN1B1-CDG patients, which was characterized as a type II congenital disorder of glycosylation (CDG), disrupting not only protein N-glycosylation but also general Golgi morphology. They found an increase of sialyl Lewis X glycans on serum proteins of all patients. Dotz et al. (2018) measured N-glycans in plasma samples of 1583 type 2 diabetes cases and 728 controls using a high-throughput MALDI-TOF MS method. Associations were investigated with logistic regression and adjusted for age, sex, body mass index, high-density lipoprotein-cholesterol, non-high-density lipoprotein-cholesterol, and smoking. Findings were replicated in a nested replication cohort of 232 cases and 108 controls. Eighteen glycosylation features were found to be significantly related with type 2 diabetes. In diabetes, fucosylation and bisection of di-antennary glycans were found to be decreased, but total and, specifically, $\alpha 2,6$-linked sialylations were increased.

6.3.1.2 Urine

Urine is another rich source of potential biomarkers. Guo et al. (2015) enriched the urinary glycoproteins from normal controls, normoalbuminuria, microalbuminuria, and macroalbuminuria patients by concanavalin A, followed by 2DLC-MS/MS analysis for relative and absolute quantification to discover the differential proteins related to type 2 diabetic nephropathy (DN). They identified a total of 478 proteins including

408 N-glycoproteins, among which there were 72, 107, and 123 differential proteins in normoalbuminuria, microalbuminuria, and macroalbuminuria, respectively. α-1-antitrypsin and ceruloplasmin showed excellent AUC values at 0.929 and 1.000, respectively, when they were used to distinguish the microalbuminuria and normoalbuminuria. Kawahara et al. (2018) quantified 729 N-glycoproteins containing 1310 unique N-glycosylation sites and 954 unique intact N-glycopeptides in urine samples collected from prostate cancer (PCa) patients and benign prostatic hyperplasia (BPH) patients (n = 6), and found a panel of 56 intact N-glycopeptides that could discriminate PCa and BPH (ROC: AUC = 1). Jia et al. (2016) identified a total of 2923 unique glycosite-containing peptides from urine samples of PCa patients with different Gleason scores by using solid-phase extraction and LC-MS/MS, and found that the majority of aggressive PCa-associated glycoproteins were more readily detected in patient's urine than serum samples. Therefore, they thought that urine could provide a potential source for biomarker discovery in patients with AG PCa. Belczacka et al. (2019) collected urine samples from 238 normal subjects and 969 patients with five different cancer types (bladder, prostate, and pancreatic cancers, cholangiocarcinoma, and renal cell carcinoma) and investigated their intact glycopeptides. They identified a total of 23 intact N-glycopeptides in the urinary samples from normal subjects, and 5 N-glycopeptides significantly and differentially expressed among the different cancer types, comparing to control samples.

6.3.1.3 Saliva

Saliva is regarded as a window to body health (Greabu et al. 2009), and increasingly used for biomarker discovery due to its easy accessibility, noninvasive collection and potential clinical value. Wu et al. (2020) identified 28, 32, and 49 endogenous glycopeptides from three normal saliva samples, and 27, 39, and 40 endogenous glycopeptides from three patients' saliva samples. However, further comparison wasn't performed and the used samples in their study were very few. Qin et al. (2017a) investigated N-linked glycome in saliva from 200 subjects (50 healthy volunteers (HV), 40 HBV-infected patients (HB), 50 cirrhosis patients (HC), and 60 HCC patients) using MALDI-TOF/TOF MS. A total of 40, 47, 29, and 33 N-glycan peaks were identified in HV, HB, HC, and HCC groups, respectively, among which 3, 2, 5, and 3 N-glycan peaks were unique in HV, HB, HC, and HCC group, respectively. They found that the abundance of fucosylated N-glycans was apparently increased in the HCC group than in any other group; however, the sialylated N-glycans were down-regulated in HCC group.

6.3.1.4 Cerebrospinal Fluids (CSFs)

Wang et al. (2016a) investigated putative molecular dynamic changes in cerebrospinal fluids collected from individuals with mild cognitive impairment (MCI) and Alzheimer's disease (AD) (n = 8) as compared to healthy controls (n = 4). Glycoproteins were enriched by lectin affinity chromatography, digested by trypsin, followed by LC-MS/MS analysis for identification and label-free quantitation. They identified 795 proteins in this study and found that there were 15 proteins (e.g., transthyretin) differentially expressed among the three groups. Schedin-Weiss et al. (2020) observed an increase in N-glycans containing bisecting N-acetylglucosamine in AD compared with nondemented controls, which could specifically bind to lectin PHA-E, as well as significantly increased binding to the lectin PHA-E in MCI and AD compared to subjective cognitive impairment (SCI) after further analysis of CSF from 242 patients. In addition, PHA-E binding correlated with CSF levels of phosphorylated tau and total tau, which was most prominent in the SCI group (R = 0.53-0.54). Palmigiano et al. (2016) compared glycosylation patterns in patients with AD (n = 24) and MCI (n = 11), as well as in healthy controls (n = 21). They found that compared with normal controls, AD patients had increased levels of bisecting-GlcNAc-type N-glycans and decreased levels of overall sialylation.

6.3.2 Tissues

Tissue specimen is a type of invaluable resource for clinical proteomic research. Although composition of tissue samples is more complex, the obtained tissues are rich in pathological information. In addition, the diseased tissues (i.e., tumour) and its matched adjacent controls can be acquired at the same time, which can provide more comparable results about differences of protein expression and its PTMs. In the existing researches, both fresh tissues and formalin-fixed paraffin-embedded (FFPE) tissues are commonly used, especially FFPE tissues are the common form of sample preservation in clinical pathology. Hinneburg et al. (2017) found that as few as 2000 cells isolated from FFPE tissue sections by laser capture microdissection were sufficient for in-depth histopathology-glycomics using porous graphitized carbon nano-LC-ESI-MS/MS. In addition, N- and O-glycan profiles were similar between unstained and haematoxylin and eosin stained FFPE samples but differed slightly compared with fresh tissues.

Cheng et al. (2016) observed pronounced differences of the N-glycosylation patterns and fucosylated N-glycans between the adjacent and liver tumour tissues from 27 HCC patients. They reported that fucosylated N-glycans and FUT8 levels could be used as markers for evaluating HCC progression. Similarly, Qin et al. (2019) analyzed the differential glycoproteomes of HCC and adjacent normal human liver tissues, and observed several proteins (such as MXRA5 and CP) with significantly changed glycosites and site-specific glycoforms. In their study, both of the HCC and adjacent human liver tissues were pooled from the ten patients, respectively. Chen et al. (2017a) quantitatively compared N-glycans from FFPE tissue slides among different pathological grades of EOC and healthy controls using MALDI-TOF MS. They observed higher expression levels of high-mannose type in EOC samples than that in healthy controls, accompanied by reduced levels of hybrid-type glycans. A combined panel composed of four high-mannose and three fucosylated neutral complex N-glycans allowed for good discrimination of EOC from healthy controls. Furthermore, abundance of 14 N-glycans (2 high-mannose-type, 2 fucosylated and sialylated complex structures, and ten fucosylated neutral complex N-glycans) exhibited specific changes across EOC grades. Li et al. (2017) studied the overall variation of glycoproteome and site occupancy of glycoproteins in high-grade ovarian serous carcinoma and benign epithelial ovarian tumour serous cystadenoma. Glycopeptides were enriched using hydrazide chemistry and labelled with an iTRAQ reagent, and analyzed by LC-MS. A total of 36 glycosite-containing peptides showed significant changes between ovarian tumour and benign tumour, while the site occupancies of ten glycopeptides increased. Wang and Tian (2020) found that 838 intact N-glycopeptides were differentially expressed between human gastric cancer and adjacent tissues, where 220 were down-regulated and 618 were up-regulated with the criteria of ≥1.5-fold change and $p<0.05$. Sethi et al. (2015) performed quantitative N-glycomes of paired tumourigenic and adjacent non-tumourigenic tissue samples from five male patients suffering from adenocarcinoma CRC. An over-representation of high mannose, hybrid, and paucimannosidic type N-glycans and an under-representation of complex N-glycans were observed as CRC-specific N-glycosylation phenotypes. Such CRC stage-specific N-glycosylation as high α2,3-sialylation and low bisecting β1,4-GlcNAcylation and Lewis-type fucosylation in mid-late relative to early stage CRC was detected. High bisecting β1,4-GlcNAcylation and low α2,3-sialylation relative to EGFR-negative CRC tissues were identified as EGFR-specific N-glycan signatures. Yang et al. (2017) quantitatively identified over 6000 global proteins and 480 glycoproteins in both primary lung squamous cell carcinoma (SqCC) and adenocarcinoma (ADC). Two glycoproteins (ELANE and IGFBP3) were identified to be only increased in SqCC, and six glycoproteins (ACAN, LAMC2, THBS1, LTBP1, PSAP, and COL1A2) were increased in ADC. Wang et al. (2018) identified 58 N-glycan compositions from lung adenocarcinoma FFPE tissue samples and found that high-mannose type and sialylated N-glycans could be used to distinguish lung adenocarcinoma and control tissue by receiver operating characteristic curve analysis. Saarinen et al. (2018) found differences in N-glycan profiles, especially the neutral glycans, between pseudomyxoma peritonei (a subtype of mucinous adenocarcinoma) tissues and the normal appendix controls, among which increased fucosylation was the most prominent alteration. Al-wajeeh et al. (2017) examined differentially expressed glycoprotein profiles of 48 pairs of breast cancer and its adjacent tissues among breast cancer women in Malaysia. In total, 11 glycoproteins were detected only in tumour tissues, but not in normal ones. Leijon et al. (2017) identified differences in N-glycomic profiles of primary metastasized and non-metastasized pheochromocytomas and paragangliomas from FFPE tissue samples. Such four groups of neutral N-glycan signals as complex-type N-glycan signals of cancer-associated terminal N-acetylglucosamine, multifucosylated glycans (complex fucosylation), hybrid-type N-glycans, and fucosylated paucimannose-type N-glycans were found to be more abundant in metastasized tumours than in non-metastasized tumours. Three groups of acidic N-glycans, multifucosylated glycans, acid ester-modified (sulfated or phosphorylated) glycans, and hybrid-type/monoantennary N-glycans, were more abundant in metastasized tumours. Fucosylation and complex fucosylation were found to be significantly increased in metastasized paragangliomas and pheochromocytomas than in non-metastasized tumours for individual tests; however, they were over the false positivity critical rate when adjusted for multiplicity testing. Möginger et al. (2018) investigated the healthy human skin N-glycome and its changes associated with basal cell carcinoma (BCC) and squamous cell carcinoma (SCC), in which matched patient samples were obtained from frozen biopsy and FFPE tissue samples. They found that complex type N-glycans exhibiting almost similar levels of α2,3 and α2,6 sialylation dominated in the human skin N-glycome. Oligomannose N-glycan levels were found to be elevated in BCC and SCC, while α2,3 sialylation levels were decreased in SCC but not in BCC.

To adopt metabolic labelling techniques, tissue culture associated methods should be a good solution. For example, Spiciarich et al. (2017) reported a glycoproteomic platform to human tissues cultured ex vivo. Both normal and cancerous prostate tissues were sliced. Peracetylated N-azidoacetylmannosamine was then used in sliced normal and cancerous prostate tissue cultures in order to add the azide groups to the sialic acid residues on cell surface and secreted glycoproteins. Therefore, the labelled sialoglycoproteins could be reacted with a biotin alkyne probe, captured with avidin resin, digested with trypsin, and analyzed by LC-MS/MS.

Of course, clinical tissue specimens are relatively rare; therefore, we can first perform experiments with cell lines and then clinical specimens are used for validation or further experiments. For example, Shan et al. (2020) found that the composition profiling of fucosylated N-glycans differed between high metastatic C8161 and low metastatic A375P cells by MALDI-TOF MS analysis, and fucosyltransferase-4 (FUT4) expression was significantly increased in C8161 cells. Then, they further demonstrated that FUT4 was overexpressed in metastatic samples by melanoma tissue arrays, and altered FUT4 expression was accompanied by a change in the migration and invasion capacity of the cells. Moreover, due to its rich information, a batch of specimens could be used comprehensively by using some versatile methods to save clinical samples. For instance, immobilized metal affinity chromatography technology for the enrichment of phosphopeptides can coenrich sialoglycopeptides, allowing for a large-scale analysis of sialoglycopeptides in conjunction with the analysis of phosphopeptides. By using this method, Hu et al. (2018) revealed a large number of previously unidentified N-linked glycopeptides after they reanalyzed the global proteomic data from breast cancer xenograft tissues.

MALDI MS imaging allows direct characterization of the spatial distribution of various molecules on tissues such as protein/peptides, N-glycans, and so on. To characterize N-glycans, PNGase F is sprayed on the sample (fresh or FFPE tissues) to release N-glycans directly on the tissues mounted on glass slides, and the samples can be analyzed followed by adding MALDI matrix using this method. Drake et al. (2020) assessed N-glycosylation patterns and compositional differences between tumour and nontumour regions of formalin-fixed clinical clear cell renal cell carcinoma (ccRCC) specimens and tissue microarrays, as well as N-linked glycan-based distinctions between cortex, medullar, glomeruli, and proximal tubule features in regions of normal kidney tissue samples. Powers et al. (2015) observed distinct differences in glycan structure between HCC versus normal tissue samples. West et al. (2018) compared the N-linked glycans in 138 HCC tissue with either adjacent untransformed or tissue from patients with liver cirrhosis but no cancer. Ten glycans, which were with increased levels of fucosylation and/or with increased levels of branching, were found to be significantly elevated in HCC tissues compared to cirrhosis or adjacent tissue. Additionally, high levels of fucosylated glycoforms were associated with shorter survival time. Holst et al. (2016) detected N-glycans from colon carcinoma and leiomyosarcoma FFPE tissues. They found that high-mannose type glycans could be attributed to various tumour regions of the tissues, while sialylated glycans distinguished different collagen-rich (tumour) regions, stroma, and others. N-glycans with only α2,3-linked sialic acids correlated mainly with stroma, tumour, and necrotic cell regions, while α2,6-linked sialic acid containing N-glycans localized with tissue morphologies attributed to necrotic tissue, collagen-rich areas, and red blood cells. Scott et al. (2019) found that the structures of the N-glycans present in regions of tumour necrosis in breast, thyroid, cervix, and liver cancer tissues were generally non-fucosylated bi-antennary or tri-antennary structures with sialic acid modifications using MALDI-FTICR MS imaging.

6.3.3 Specific Biomarkers

6.3.3.1 Igs

As an important PTM, glycosylation significantly affects the structure and function of immunoglobulins, and even subtle differences can reflect different physiological and pathological states. Therefore, the aberrant glycosylation of Igs can be used as various disease biomarkers.

6.3.3.1.1 Cancers

Kazuno et al. (2016) analyzed glycosylation status of serum IgG from patients with prostate diseases and normal subjects, and found that an absence of the terminal hexose of N-linked glycans was closely connected to the progression of PCa. For OVC, Ruhaak et al. (2016b) quantified concentrations of IgG, IgA, and IgM, as well as glycosylation profiles in serum samples from women diagnosed with EOC (n = 84) and matched healthy controls (n = 84) using MRM MS. Multiple glycopeptides from IgA, IgG, and IgM were found to be differentially expressed in serum of EOC patients compared with controls. In particular, IgG-specific glycosylation profiles were the most powerful in discriminating between EOC patients and controls. A biomarker panel containing 11 glycoproteins was discovered to increase the accuracy of OVC prediction when combined with CA-125, the most widely used biomarker for OVC. Miyamoto et al. (2018) also performed the protein- and site-specific quantitation of such serum proteins as IgA, IgG, IgM, α-1-antitrypsin, transferrin, α-2-macroglobulin, haptoglobin, α1-acid glycoprotein, and complement C3 using a MRM-based method, in which the training set consisted of 40 cases and 40 controls, while the independent test set included 44 cases and 44 controls. Gebrehiwot et al. (2019) identified significantly up-regulated 35 N-glycans in the sera of all breast cancer (BC) patients compared to the normal controls (NC), in which 17 complex type N-glycans showed strong diagnostic potential (AUC = 0.8-1) for the early stage (I and II) BC patients. Most of these glycans contained core-fucosylated, multiply branched, and sialylated structures, and their abundance was strongly associated with greater invasive and metastatic potential of cancer. Two core-fucosylated and agalactosylated IgG glycans (m/z 1591 and 1794) could specifically distinguish ($p \leq 0.001$, AUC = 0.944 and 0.921, respectively) stage II patients from NC. Kawaguchi-Sakita et al. (2016) compared N-glycans of serum IgG Fc region in 90 BC patients and 54 cancer-free controls using MALDI MS. A characteristic pattern of IgG Fc region N-glycosylation was identified in BC patients, which could distinguish between BC patients and controls using a multiple logistic regression model. They found that serum IgG N-glycan structures at stage 0 breast cancer patients had already been different from that of normal controls. For liver diseases, Yuan et al. (2015b) found HCC patients had increased circulating IgG1, IgG3, IgA1, and IgM compared to healthy controls, while had a significantly higher concentration of IgG1 and IgM but lower IgG2 concentration compared to cirrhosis patients. In addition, galactose-deficient core-fucosylated glycoforms plasma IgG subclasses 1-4 increased in cirrhosis and HCC patients. Such glycoforms as FA2G0 and FA2B0 increased in all IgG subclasses, whereas FA2G2 decreased.

Ho et al. (2015) compared serum IgG Fc glycosylation profiles in 76 patients with HBV-related liver cirrhosis and 115 patients with chronic hepatitis B (CHB) before and after 48 weeks of anti-HBV nucleos(t)ide analogue treatment using high-throughput LC-MS, as well as the glycosylation profiles in 108 healthy controls. Higher aberrant serum IgG Fc glycosylation, particularly galactose deficiency, was found in patients with CHB and with cirrhosis than in healthy controls, while the IgG galactose deficiency was correlated with the severity of liver necroinflammation and fibrosis in CHB. Multivariate logistic regression analyses showed that the IgG Fc glycoform with fucosylation and fully galactosylation was associated with a total Knodell necroinflammation score of ≥ 7 (odds ratio, 0.74; 95% CI, 0.56–0.97) and an Ishak fibrosis score of ≥ 3 (odds ratio, 0.69; 95% CI, 0.49–0.97). Administration of antiviral therapy for 48 weeks could reverse aberrant IgG Fc glycosylation in patients with CHB from week 12 onward, but could not reverse glycosylation in patients with cirrhosis. Aberrant serum IgG Fc glycosylation in patients with CHB affected IgG opsonizing activity and could be reversed by antiviral therapy. Liu et al. (2018c) compared the IgG N-glycome profiling of colorectal benign patients, CRC, and normal individuals, and found nine differentially expressed glycans in disease groups compared with controls, among which five were significantly changed in CRC patients at all tumour node metastasis stages as compared with controls, and core-fucosylation, sialylation, and sialo core-fucosylation were found to be possibly associated with CRC progression. Liu et al. (2020) studied the association of IgG isomeric glycosylation with CRC, and identified 28 IgG glycans and 79 compositional isomers by PGC-based nano-LC-ESI-MS/MS analysis. They found that CRC was associated with the increase of IgG agalactosylation, decrease of IgG sialylation and fucosylation of sialylated glycans. These three compositional isomers, $H_3N_4F_1$-a, b, and $H_4N_3S_1F_1$-e, could distinguish patients with CRC and early stages from controls. Theodoratou et al. (2016) analyzed plasma IgG glycans in 1229 CRC patients and correlated with survival outcomes. They found that decrease in galactosylation and sialylation (of fucosylated IgG glycan structures), as well as increase in bisecting GlcNAc in IgG glycan structures, were strongly associated with all-cause and CRC mortality. Zou et al. (2020) compared IgG Fc N-glycopeptides in the plasma from 46 CRC patients and 67 healthy controls, and found 11 N-glycopeptides (e.g., IgG1 G0N) were significantly and differentially expressed in CRC patients, among which abundance of N-acteylglucosamine was increased, while levels of galactosylation, fucosylation, and sialylation was decreased. Zhang et al. (2016) analyzed circulating disease-specific IgG (DSIgG) glycopeptides attached to IgG Fc region at the site of Asn297 from 846 serum samples of 443 patients with benign gastric diseases (BGDs) and 403 patients with gastric cancer. They found that DSIgG1 G1S, DSIgG2 G0F, G1, G2F, and G2FS as well as DSIgG2 galactosylation and sialylation were significantly associated with sex in BGD patients, while the age-specific glycosylation features from DSIgG between BGD and gastric cancer patients had similar change trends. Galactosylation, sialylation, and bisecting GlcNAc from DSIgG were also found to have significant changes between BGD and gastric cancer patients. The ratio of G2FN/G1FN (from DSIgG2) could be used to distinguish female BGD patients from female gastric cancer patients over the age range of 20–79 years, with 82.6% sensitivity, 82.6% specificity, and 0.872 AUC. Ruhaak et al. (2015) observed that there were eight glycans in IgG differed significantly between non-atrophic gastritis (NAG) and gastric cancer, three glycans distinguished NAG from duodenal ulcer (DU), and eight glycans differed between DU and gastric cancer. Compared with the DU and NAG, there was lower sialylation in gastric cancer patients.

6.3.3.1.2 Immune and Inflammatory Diseases

Wuhrer et al. (2015) analyzed IgG Fc glycosylation, including sialylation, galactosylation, fucosylation, and bisecting GlcNAc, in 48 paired CSF and serum samples from adult patients with multiple sclerosis or a first demyelinating event highly suggestive of multiple sclerosis (designated as multiple sclerosis cases), and from healthy volunteers and patients with other non-inflammatory diseases (control group). They found that IgG1 glycosylation patterns were different in CSF vs serum. In multiple sclerosis patients vs controls, bisecting GlcNAc of IgG1 were elevated and afucosylation and galactosylation were reduced in CSF, but not in serum. Song et al. (2018) analyzed glycosylation of whole serum and site-specific glycosylation of immunoglobulins in 27 serum samples from steroid-induced osteonecrosis of the femoral head (ONFH) patients and 25 from gender and age-matched controls using ESI-Q-TOF and ESI-Triple-Quadruple via MRM. They found that the increased non-sialylated and non-fucosylated N-glycans, as well as the decreased fucosylated N-glycans, were associated with the development of ONFH. Yuan et al. (2015a) investigated the differential expression of specific glycosylation patterns of anti-thyroglobulin IgG (TgAb IgG) from Hashimoto's thyroiditis patients and healthy blood donors, and found TgAb IgG in patients exhibited higher glycosylation levels than those in healthy controls. Huang et al. (2017c) found that galactosylation and sialylation of IgG N-glycans in RA cases were significantly lower than that in healthy subjects. Harre et al. (2015) demonstrated that desialylated, but not sialylated, immune complexes could enhance osteoclastogenesis in vitro and in vivo. In addition, the Fc sialylation state of random IgG and specific IgG autoantibodies determined bone architecture in patients with RA. Rombouts et al. (2015a) found that in the analyzed 24 patient samples with RA, the vast majority of anti-citrullinated protein antibodies (ACPA)-IgG molecules exhibited a higher molecular weight due to the presence of N-glycans in the ACPA variable domains in comparison with other autoantibodies or non-autoreactive IgG. Wang et al. (2017) identified 20 sulfated and 4 acetylated N-glycans in human serum IgGs, and further quantified these N-glycans using MRM method from 277 patients with RA and 141 healthy individuals to identify N-glycan biomarkers for the classification of both rheumatoid factor -positive and negative RA patients, as well as ACPA-positive and negative RA patients. Rombouts et al. (2015b) collected serum samples of patients with ACPA-positive arthralgia (n = 183) at baseline and at various time points of follow-up, analyzed ACPA-IgG1-Fc glycopeptides and compared to those of total IgG1 by nano-liquid chromatography mass spectrometry. They found that ACPA displayed significant changes in Fc galactosylation and fucosylation prior to the onset of RA. ACPA-IgG1 and total IgG1 from arthralgia patients displayed similar Fc glycosylation patterns at baseline. However, ACPA exhibited a decrease in galactose residues in RA patients at the onset of arthritis, but not in UA patients. Galactosylation of total IgG1 was also decreased in RA at the onset of arthritis. In addition, there was a higher degree of ACPA-IgG1 Fc core fucosylation at baseline than that of total IgG1, and it could further increase prior to diagnosis. Hafkenscheid et al. (2017) analyzed the N-linked glycosylation of total, Fc and $F(ab')_2$ fragments, as well as heavy and light chains of ACPA-IgG and noncitrulline-specific (control) IgG from plasma and/or synovial fluid of nine ACPA positive RA patients by UHPLC and MALDI-TOF MS. The pattern and level of Fab-glycosylation were found to differ markedly between ACPA-IgG and noncitrulline-specific IgG as well as between ACPA isolated from the synovial fluid or blood. ACPA-IgG molecules exhibited highly sialylated glycans in their Fab domain, while Fab-linked glycans were estimated to be present on over 90% of ACPA-IgG, which was five times higher than that of control IgG. This feature was more prominent on ACPA isolated from synovial fluid compared with peripheral blood. Sonneveld et al. (2017) observed significantly decreased galactosylation and sialylation levels in total IgG1 purified from plasma of patients with red blood cells (RBC)-bound antibodies compared to healthy controls. And the anti-RBC-autoantibodies contained even lower galactosylation, but higher sialylation and lower bisection levels. Compared with alloantibodies against RBCs, RBC-bound IgG1 Fc-fucosylation was not different between patients and healthy controls. Chen et al. (2020a) investigated the differences of IgA concentrations and glycosylation profiles between patients with kidney diseases and healthy controls, in which 30 clinical samples were included, including samples from 8 patients diagnosed with IgA nephropathy (IgAN), 16 patients with non-IgAN chronic kidney diseases, and 6 healthy controls. They found that the profiles of IgA and IgA-glycopeptides were different in patients with chronic kidney diseases and IgA nephropathy compared with healthy controls. There was the highest IgA1 concentration in IgAN, followed by CKD and healthy controls. The distribution of IgA1 concentrations was more diverse in patients with kidney disease than healthy controls, but the distribution of IgA2 concentrations varied more in CKD patients only. Although glycopeptides varied between CKD and control and between IgAN and healthy controls, there was no clear separation between the IgAN and CKD groups, or among the three groups. Plomp et al. (2017) investigated the IgG glycosylation in relation to measurements of inflammation, such as plasma levels of C-reactive protein and interleukin-6, and of metabolism. The plasma Fc glycosylation profiles of IgG1, IgG2, and IgG4 in a cohort of 1826 individuals were analyzed by LC-MS. For all subclasses, decreased galactosylation and sialylation and increased core fucosylation associated with poor metabolic health, i.e., increased inflammation, low serum high-density lipoprotein cholesterol, and high triglycerides. Compared to IgG1 and IgG4, IgG2 consistently showed weaker associations of its galactosylation and sialylation with the metabolic markers, but the trend of the associations was overall similar for all IgG subclasses. Šimurina et al. (2018) compared IgG Fc glycosylation of 3441 plasma samples obtained from 2 independent cohorts of patients with Crohn's disease (CD) (874 from Italy and 391 from the US) or ulcerative colitis (UC) (1056 from Italy and 253 from the US and healthy controls; 427 in Italy and 440 from the US) by LC-MS analysis. They found that the levels of IgG galactosylation were lower in patients with CD or UC than in controls. And the decreased galactosylation associated with more severe CD or UC, including the need for surgery in patients with UC or with CD vs controls. IgG fucosylation was increased in patients with CD vs controls, but decreased in patients with UC vs controls. The number of samples used in this study is quite large, which might probably increase our understanding of mechanisms of CD and UC pathogenesis, as well as their diagnostics or guide treatment.

6.3.3.1.3 Other Diseases

Russell et al. (2017) compared plasma IgG N-glycans between 94 patients with Parkinson's disease and 102 controls, and found that 7 glycan peaks and 11 derived traits had statistically significant differences. Out of the 7 glycan peaks, 4 (GP5, GP8, GP17, and GP20) were selected as the potential biomarker with a sensitivity of 87.2% and a specificity of 92.2%. Yu et al. (2016) reported various patterns of changes in IgG glycosylation associated with age by analyzing IgG glycosylation in 701 community-based Han Chinese (244 males, 457 females; 23–68 years old). Such 11 IgG glycans as FA2B, A2G1, FA2[6]G1, FA2[3]G1, FA2[6]BG1, FA2[3]BG1, A2G2, A2BG2, FA2G2, FA2G2S1, and FA2G2S2 were found to change considerably with age. Specific combinations of these glycan features could explain 23.3% to 45.4% of the variance in chronological age in this population. To search IgG N-glycans as potential hypertension biomarker in the Kazakh population, Gao et al. (2017) analyzed plasma samples of 150 Kazakh (52 with hypertension, 23 with prehypertension, 75 with normal blood pressure) using UPLC-ESI-Q-TOF MS. Fourteen IgG subclass-specific Fc N-glycopeptide structures, as well as one derived glycosylation trait in subclasses IgG2/3 and IgG4, were found to correlate with systolic blood pressure and/or diastolic blood pressure. Liu et al. (2018b) compared subclass-specific IgG Fc glycosylation profile in 274 hypertensive patients and 356 healthy controls from the four northwestern Chinese minority populations (Uygur, Kazak, Kirgiz, and Tajik). They found that ten IgG N-glycan traits (i.e., IgG1G0F, IgG2G0F, IgG2G1FN, IgG2G1FS, IgG2G2S, IgG4G0F, IgG4G1FS, IgG4G1S, IgG4G2FS, and IgG4G2N) were significantly associated with hypertension across the four ethnic groups. Harper et al. (2015) reported on the relative galactosylation, sialylation, fucosylation, and bisection of total IgG1 and IgG2 and anti-proteinase-3 (anti-PR3) specific IgG1 from patients with granulomatosis with polyangiitis (GPA), as well as their correlations with inflammatory cytokines and disease activity and progression. They found that the IgG1 and IgG2 galactosylation, sialylation, and bisection were reduced in GPA patients compared to in healthy controls. Anti-PR3 IgG1 ANCA Fc galactosylation, sialylation, and bisection were reduced compared to total IgG1 in GPA patients, among which galactosylation correlated with inflammatory cytokines and time to remission but not Birmingham vasculitis activity scores. Wu et al. (2016c) compared the site-specific IgE glycosylation from a patient with a novel hyper-IgE syndrome linked to mutations in phosphoglucomutase-3 (PGM3), which was an enzyme involved in synthesizing UDP-GlcNAc, a sugar donor widely required for glycosylation, and a patient with atopic dermatitis as a control subject. However, there were no significant differences found between the two IgE samples. Therefore, they thought that despite alterations occurring in the N-glycome of immune cells from patients with PGM3 mutations, the elevated IgE in allergy and hyper-IgE syndrome might not be related to glycosylation on the antibody itself. Of course, the sample numbers should be further added to obtain more reliable information.

6.3.3.2 Prostate-Specific Antigen

The concentration of PSA in serum is used as an early detection indicator of PCa, however, suffering from limited specificity. Therefore, researchers have gradually moved toward analyzing its glycosylation instead of only its protein concentration in the bodily fluids. Haga et al. (2019) explored PCa-specific glycosylation subtypes of PSA and found that abundance of multisialylated LacdiNAc (GalNAcβ1,4GlcNAc) structures was significantly up-regulated in the PCa group compared to the BPH group. Kammeijer et al. (2018) identified a total of 67 N-glycopeptides from the PSA pooled from patients and found low levels of α2,3-sialylated glycans in healthy volunteers when compared with the patient profiles. Jia et al. (2017) investigated the PSA glycan profiles in Chinese population after they purified PSA from expressed prostate secretions (EPS)-urine samples from 32 BPH and 30 PCa patients. They found that most of the PSA glycans from EPS-urine samples were complex type bi-antennary glycans. And a significant increase of glycan FA2 and FM5A2G2S1 in PCa was found compared with BPH patients, but overall, there was no distinct difference of PSA glycans between BPH and PCa patients.

6.3.3.3 Haptoglobin

Huang et al. (2017b) found the glycan isoform patterns of serum haptoglobin could effectively distinguish early stage HCC from cirrhotic patients, among which the statistically significant glycan isomers were either branch fucosylated or composed of α2,6-linked sialic acid moieties. To analyze a large cohort of clinical serum samples, Zhu et al. (2015) developed a method based on a 96-well plate platform to evaluate fucosylation changes of serum haptoglobin between HCC vs cirrhosis, in which only 10 μL of serum was required for glycan extraction and processing for MALDI-QIT MS analysis. They found that the bifucosylated tetra-anntenary glycan was up-regulated in HCC samples of all aetiologies using this method. Zhang et al. (2015) found that the fucosylated glycans of haptoglobin were up-regulated in the case of HCC serum samples vs cirrhosis samples. Zhu et al. (2019) compared the microheterogeneity of site-specific intact N-glycopeptides of serum haptoglobin between early HCC and liver cirrhosis. In total, 93, 87, and 68 site-specific N-glycopeptides were identified in early HCC, liver cirrhosis, and healthy controls, respectively. They found that increased branching with hyper-fucosylation and sialylation increased the variety of N-glycopeptides in liver diseases compared to healthy controls. And five site-specific N-glycopeptides on sites N184 and N241 were significantly up-regulated in early HCC compared to cirrhosis (p<0.05) and normal controls (p≤0.001). Kim et al. (2017b) found that N-glycan variation of serum haptoglobin were associated with patients with gastric cancer after comparing N-glycans in serum samples from patients with gastric cancer (n = 44) and healthy control (n = 44). The AUC of their established frequency marker (*f*-marker) panels based on the tendency of high N-glycan expression to conclude patients and control group were 0.82 and 0.79, respectively. Lee et al. (2018a) characterized 96 glycopeptides of serum haptoglobin across all gastric cancer and healthy control samples, where 3 glycopeptides exhibited exceptionally high control-to-cancer fold changes along with receiver operating characteristic curve areas of 1.0. Takahashi et al. (2016) undertook site-specific analyses of N-glycans on haptoglobin in the sera of patients with five types of operable gastroenterological cancer (oesophageal, gastric, colon, gallbladder, pancreatic), a non-gastroenterological cancer (PCa) and normal controls. They found that monofucosylated N-glycans were significantly increased at all glycosylation sites (Asn184, Asn207, Asn211, and Asn241) in all the above cancer samples, while difucosylated N-glycans at Asn184, Asn207, and Asn241 were detected only in cancer samples. Both Lewis-type and core-type fucosylated N-glycans were increased in gastroenterological cancer samples, but only core-type fucosylated N-glycan was relatively increased in PCa samples. In metastatic PCa, Lewis-type fucosylated N-glycan was also increased. Wu et al. (2018) established a MS approach to systematically dissect the microheterogeneity of two important serum proteins (α1-acid glycoprotein and haptoglobin) and relate glycan features to drug- and protein-binding interaction kinetics. They found that the degree of N-glycan branching and extent of terminal fucosylation could attenuate or enhance these interactions, providing important insight into drug transport in plasma.

6.3.3.4 Other Glycoproteins

Transferrin is an 80 kDa glycoprotein and the glycoform at two N-glycosylation sites is comprised of a disialylated bi-antennary oligosaccharide as the major form and minor species with fucosylated or triantennary structures. Wada (2016) analyzed native transferrin for rapid CDG screening by MS. They found immature glycoforms even existed in healthy individuals and the diagnosis of CDG based on molecular phenotypes required quantitative evaluation. van Scherpenzeel et al. (2015) compared protein-specific glycoprofiling of intact transferrin in the CDG by analyzing plasma samples from controls (n = 56), patients with known defects (n = 30), and patients with secondary (n = 6) or unsolved (n = 3) cause of abnormal glycosylation. Nigjeh et al. (2017) found that the level of N-glycosylated peptides derived from galectin-3-binding proteins (LGALS3BP) were frequently elevated in plasma from pancreatic ductal adenocarcinoma patients, which was consistent with that in the tumour tissue. Kim et al. (2019) measured glycosylation alteration in AFP, vitronectin (VTN), and α-1-antichymotrypsin from in HCC, cirrhosis, and healthy sera using parallel reaction monitoring analysis. They found that the AUC (0.944) for the combination of AFP and VTN increased more than for a single glycopeptide (AUC = 0.889 for AFP and 0.792 for VTN) with respect to discriminating between HCC and cirrhosis serum.

Darebna et al. (2017) compared N-glycosylation of serum haptoglobin, hemopexin, and complement factor H in HCC, CRC, and liver metastasis of CRC. They observed significant increase in fucosylation of all three proteins in HCC of hepatitis C viral aetiology. Togayachi et al. (2017) examined the expression, secretion, and glycosylation of Secretogranin III

(SgIII) and found that SgIII existed in all types of lung cancer, while low-molecular-weight SgIII (short-form SgIII) was specifically found in small cell lung carcinoma (SCLC) culture medium. They thought that fucosylated short-form SgIII might be a valuable biomarker for SCLC. Gbormittah et al. (2015) characterized the occupancy and degree of heterogeneity of individual N-glycosylation sites of clusterin in the plasma of patients diagnosed with localized ccRCC, before and after curative nephrectomy (n = 40). They determined the levels of targeted clusterin glycoforms containing either a bi-antennary digalactosylated disialylated (A2G2S2) glycan or a core-fucosylated bi-antennary digalactosylated disialylated (FA2G2S2) glycan at N-glycosite N374, and found that the presence of these two clusterin glycoforms differed significantly in the plasma of patients prior to and after curative nephrectomy for localized ccRCC. Removal of ccRCC led to a significant increase in the levels of both FA2G2S2 and A2G2S2 glycans in plasma clusterin. Losev et al. (2019) found that N-glycans attached to human tau could affect its aggregation. Balmaña et al. (2015) analyzed the sialyl-Lewis X levels on ceruloplasmin from sera of pancreatic adenocarcinoma (PDAC) (n = 20), chronic pancreatitis (n = 14), and healthy controls (n = 13) and found that sialyl-Lewis X tended to be increased in the PDAC group. Giménez et al. (2015) compared the structures of the N-linked oligosaccharides on human α1-acid-glycoprotein from patients with pancreatic cancer and patients with chronic pancreatitis. They found an increase in fucosylated tri-antennary trisialylated and fucosylated tetra-antennary trisialylated N-linked glycans in the pancreatic patients when compared to the pancreatitis patients. And the increased abundance of these glycans differed between the various stages of pancreatic cancer.

In these studies, different type of samples associated with various diseases were used and had provided valuable information for researchers, although different sample sizes were used. Generally, appropriate sample size is necessary in order to obtain credible results. In addition, adoption of suitable analytical methods is also important in order to acquire more valuable information. For example, Reiding et al. (2019b) evaluated and compared the performance of three high-throughput released N-glycome analysis methods by assessing the same panel of serum samples obtained at multiple timepoints during the pregnancies and postpartum periods of healthy women and patients with RA: hydrophilic-interaction ultra-high-performance liquid chromatography with fluorescence detection with 2-aminobenzamide (2-AB) labelling of the glycans, multiplexed capillary gel electrophoresis with laser-induced fluorescence detection with 8-aminopyrene-1,3,6-trisulfonic acid labelling, and MALDI-TOF MS with linkage-specific sialic acid esterification. They found that the methods were similar in their detection and relative quantification of serum protein N-glycosylation, but the non-MS methods showed superior repeatability over MALDI-TOF MS and allowed the best structural separation of low-complexity N-glycans, while MALDI-TOF MS achieved the highest throughput and provided compositional information on higher-complexity N-glycans.

6.4 CHARACTERIZATION OF THERAPEUTIC GLYCOPROTEINS

N-glycosylation has a profound impact on the biophysical and pharmacokinetic properties of therapeutic glycoproteins, including their conformations, structural stabilities, serum half-lives and biological efficiencies, which are related to safety and efficacy of drugs (Higel et al. 2016; Wang et al. 2020a). Yet, many factors influence the glycosylation of biotherapeutics, ranging from expression systems and cell culture processes to downstream purification and storage strategies. Therefore, identification of their glycosylation and annotation of their structures is critical for not only new drug candidates but also monitoring batch-to-batch consistency of established drugs and comparing biosimilars and biobetters to originator drugs, in order to ensure the consistent quality and clinical performance of biopharmaceuticals throughout their product lifecycle (Lamanna et al. 2018). Here, we summarize applications of MS-based methods in analysis of therapeutic glycoproteins including antibody drugs, recombinant protein drugs and vaccines.

6.4.1 Antibody Drugs

mAbs have become an important class of therapeutic drugs for multiple indications (e.g., cancers and inflammatory diseases) (Elgundi et al. 2017), and over 60 monoclonal antibody drugs have been approved by FDA (Singh et al. 2018). Glycosylation significantly impacts structure, function, and therapeutic effects of therapeutic antibodies (Cymer et al. 2018; Liu 2015). Thus, it is important to characterize glycosylation to control the quality of antibodies.

van der Burgt et al. (2019) structurally characterized two IgG1 mAbs (NIST mAb standard and trastuzumab) using ultrahigh resolution matrix-assisted laser desorption/ionization in-source decay (MALDI-ISD)-FT-ICR MS. Martín et al. (2015) used $^{12}C_6$ and $^{13}C_6$ 2-aminobenzoic acid differential labelling of N-glycans followed by ultra-performance hydrophilic interaction liquid chromatography with online fluorescence and tandem mass spectrometry to compare the glycosylation present of different batches of a commercial chimeric IgG1 mAb. Váradi et al. (2020) characterized site-specific glycosylation of cetuximab including potential glycosylation sites with their relative site occupancy using multiple fractionation methods and capillary electrophoresis MS. They found that glycans with terminal NGNA and α-Gal were located on the Fd fragment while the Fc/2 glycans were core-fucosylated bi-antennary structures with different level of β-galactosylation. Mozziconacci et al. (2016) compared the chemical stability of four IgG1 Fc glycoforms, three well-defined IgG1 Fc glycoforms (high-mannose-Fc, Man5-Fc, and N-acetylglucosamine-Fc) and a nonglycosylated Fc protein (N297Q-Fc), and found that different glycans not only affected chemical degradation differently but also led to different impurity profiles. Considering that the microheterogeneity of Asn297 glycans may impact the binding affinity of the Fc domain to FcγRIIIA, Kang et al. (2020) prepared three well-defined glycosylated and one nonglycosylated IgG4 Fc, evaluated binding affinities to FcγRIIIA of the four IgG4 Fc and correlated with physicochemical properties. Lin et al. (2015) modified the Fc-glycan structures to a homogeneous glycoform and investigated their effector activities. They found that the bi-antennary N-glycan structure with two terminal α2,6-linked sialic acids at position 297 of the Fc region was a common and optimal structure for the enhancement of antibody-dependent cell-mediated cytotoxicity, complement-dependent cytotoxicity (CDC), and anti-inflammatory activities. More et al. (2018) studied the effects of varying glycosylation on structural flexibility (especially in the CH2 domain) by hydrogen exchange-MS on the overall conformational stability, chemical stability, and receptor binding profiles of 4 IgG1 Fc glycoforms expressed and purified from *Pichia pastoris*. They found that progressively decreasing the size of the N-linked N297 glycan from high mannose (Man8-Man12), to Man5, to GlcNAc, to nonglycosylated N297Q could result in progressive increases in backbone flexibility. Glycan truncation in two potential aggregation-prone regions could significantly increase flexibility. In addition, the increased local backbone flexibility was correlated with the increased deamidation at asparagine 315, but faster oxidation of tryptophan 277 was observed to be correlated with decreased local backbone flexibility. They also observed a trend of increasing C'E glycopeptide loop flexibility with decreasing glycan size, which correlated with their FcγRIIIa receptor binding properties. Rouwendal et al. (2016) carried out a study on the production, purification, and functional evaluation of anti-HER2 IgA antibodies as anticancer agents in comparison to the anti-HER2 IgG1 trastuzumab. They compared released glycans from different IgA constructs obtained from various cell lines by MALDI-TOF MS and found profound differences in glycosylation traits across the IgA isotypes and cell lines used for production, including sialylation and linkage thereof, fucosylation (both core and antennary) and the abundance of high-mannose type species. Increases in sialylation were found to positively correlate with in vivo plasma half-lives.

Multiple biosimilar mAbs are becoming available for a single originator drug, so it is critical to demonstrate their similarity with the reference product to ensure its quality characteristics, such as physicochemical and biological properties, safety, and efficacy. Kim et al. (2017a) analyzed 103 unexpired lots in total of European Union (EU) and US Herceptin® and observed 2 sequential drifts in the sum of %afucose and %high mannose in N-glycan profiles, FcγRIIIa binding activity and antibody-dependent cellular cytotoxicity (ADCC) activity in EU and US Herceptin® lots. Lee et al. (2018b) conducted similarity assessment between CT-P10, EU-Rituximab and US-Rituximab, focusing on the physicochemical and biological quality attributes. They found that CT-P10 had identical primary and higher-order structures compared to the original product. Although CT-P10 contained the same conserved glycan species and relative proportion with the reference medicinal products, the total afucosylated glycans were slightly higher than in EU- or US-Rituximab. However, the effect of the afucosylation level in CT-P10 drug product on Fc receptor binding affinity or antibody-dependent cell-mediated cytotoxicity was negligible according to their spiking study with highly afucosylated sample. SB2 (Flixabi® and Renflexis®) is a biosimilar to Remicade® (infliximab). Hong et al. (2017) performed a characterization test with more than 80 lots of EU- and US-sourced reference product to determine whether critical quality attributes met quality standards. They found that SB2 was similar to the reference product, and although a few differences in physicochemical attributes including glycan structures existed among biosimilars and reference products, these differences might not be clinically meaningful. For example, the differences in charged glycans and charge variants between

SB2 and reference product were unlikely to affect the immunogenicity of SB2. Planinc et al. (2017) evaluated the batch-to-batch consistency of the N-glycosylation of infliximab, trastuzumab, and bevacizumab, as well as the consistency of the N-glycosylation of bevacizumab stored in polycarbonate syringes (for off-label drug use) for three months. They found that all batches of the studied therapeutic glycoproteins varied considerably (especially in galactosylation), while the N-glycosylation of bevacizumab remained unchanged during the three-month storage. Luckily, these samples with significantly different N-glycosylation profiles showed no significant variations in their biological activity, and they inferred that the differences were probably not therapeutically significant for these three proteins. Therefore, it is very necessary to establish threshold values for batch-to-batch N-glycosylation variations and regularly test batch-to-batch glycosylation consistency.

Glycosylation of antibodies can be affected by the expression systems and culture conditions during production. Therefore, Liu et al. (2017c) characterized the comprehensive glycan profiling of a biosimilar candidate of cetuximab which was produced in Chinese hamster ovary (CHO) cell lines using normal-phase high-performance liquid chromatography coupled with MALDI MS. They observed several abnormal N-linked glycans containing NeuAcLac residues in the biosimilar. Even the amount of the unusual glycans were limited, but their existence directly impacted on the accuracy of the quality control. Maier et al. (2016) analyzed heterogeneous N-glycans from various sources of IgGs including human plasma derived IgG, a CHO produced monoclonal antibody glyco-engineered for non-fucosylation, a Sp2/0 mouse myeloma cell line produced monoclonal antibody and the latter antibody additionally α2,6-sialylated with Neu5Ac. Sialic acids on human serum IgG Fc are almost exclusively α2,6-linked, while recombinant IgGs expressed in CHO cells have sialic acids through α2,3-linkages because of the lack of the α2,6-sialyltransferase gene. Zhang et al. (2019a) investigated the impact of different types of sialylation to the conformational stability of IgG through hydrogen/deuterium exchange and limited proteolysis experiments. They found that only α2,3-linked sialic acid on the 6-arm (the major sialylated glycans in CHO expressed IgG1) could destabilize the CH2 domain; however, the α2,6-linked sialic acid on the 3-arm (the major sialylated glycan in human-derived IgG), and the α2,3-linked sialic acid on the 3-arm, did not have this destabilizing effect. Kim et al. (2018) reported the production of an anti-cancer monoclonal antibody against protein CD20 from egg whites of transgenic hens, and validated the bio-functional activity of the protein in B-lymphoma and B-lymphoblast cells. They found that the chickenized CD20 monoclonal antibody (cCD20 mAb) exhibited 14 N-glycan patterns with high-mannose, afucosylation, and terminal galactosylation. The cCD20 mAb did not exhibit significantly improved Fab-binding affinity, but showed markedly enhanced Fc-related functions, including CDC and ADCC compared to commercial rituximab, a chimeric mAb against CD20. Zhou et al. (2020a) studied the effects on IgG N-glycoforms of different components in hybridoma culture media, specifically compared bovine serum albumin (BSA) with other small molecules using MALDI-QIT-TOF-MS^n-based approach. They found that the addition of macromolecular protein BSA could significantly change both glycan species and glycosylation levels of IgG, while small molecular additives (e.g., glutamine) caused little change in glycan species and levels.

To obtain comprehensive and reliable information for antibody drugs, various MS-based strategies can be adopted alone or in combination due to their different analytical principles (shown in Table 6.2).

Tran et al. (2015) employed top- and middle-down analyses with multiple fragmentation techniques including electron transfer dissociation (ETD), electron capture dissociation (ECD), and MALDI-ISD for characterization of a reference mAb IgG1 and a fusion IgG protein. Cotham and Brodbelt (2016) used a middle-down strategy that capitalized on the high energy deposition and tunability of UV photoactivation to perform the detailed primary sequence analysis and glycosylation site localization of therapeutic monoclonal antibody subunits. Giorgetti et al. (2020) analyzed seven worldwide health authorities approved mAbs (Adalimumab (huIgG1, CHO), natalizumab (hzIgG4, NS0), nivolumab (huIgG4, CHO), palivizumab (hzIgG1, SP2/0), infliximab Remicade® (chIgG1, SP2/0), rituximab (chIgG1, CHO), and trastuzumab (hzIgG1, CHO)) using a combination of bottom-up, middle-up, and intact molecule levels with a capillary electrophoresis-MS coupling, in order to get information about their charge heterogeneity, PTMs, notably major N-glycosylation forms, their location, and relative quantitation. Groves et al. (2020) described two enzymatic protocols for generating Fc glycan variants from the first mAb-based reference material (RM), RM 8761 (NISTmAb RM), and characterized both global and localized changes in higher order structure between the RM and these Fc-glycan variants using hydrogen/deuterium exchange

TABLE 6.2   Different MS Strategies/Methods and Their Characteristics

MS Strategy and Method	Characteristics	MS Strategy and Method	Characteristics
Bottom-up proteomic strategy	Intact proteins or their large fragment mixtures are first digested into small fragments of peptides followed by MS analysis, in which protein higher-order structures have been disrupted but the analytical sensitivity and sequence coverage are significantly higher due to its easier fragmentation.	Ion mobility spectrometry-mass spectrometry	It is able to separate complex mixtures of ions based on their shape and/or charge, yielding structural information complementary to molecular mass measurements, e.g. isomeric glycans.
Middle-down proteomic strategy	Proteins are digested into longer peptides (e.g. antigen-binding (Fab) and crystallizable (Fc) fragments of IgG) by limited proteolytic digestion followed by MS analysis, in which protein higher-order structures can be partially preserved.	MALDI-MS imaging	It is a two-dimensional MALDI-MS technique, which can elucidate both the spatial distribution and relative abundance of biomolecules (e.g., proteins and glycans) without extraction, purification, separation, or labelling of biological samples.
Top-down proteomic strategy	The intact proteins are directly analyzed by MS without prior digestion, which can provide more accurate and rich higher-order structure information for proteins, but is relatively insensitive and requires large amounts of samples.	Multiple reaction monitoring method	It is a targeted proteomic method and can be used for relative and absolute quantification of multiple components simultaneously in complicated mixtures (e.g. biofluids), which is highly reproducible, specific, and sensitive.
Hydrogen deuterium exchange-mass spectrometry	It can be used to obtain information on protein structure, protein-protein, or protein-ligand interaction sites, and conformational changes.	Native mass spectrometry	It has the capability to gain insights into the behaviour of intact protein, non-covalent protein-protein and protein-ligand complexes under physiological conditions, such as stoichiometry, relative or absolute binding affinities, and specificities.

MS and ion-mobility spectrometry-mass spectrometry measurements. They revealed that the decreased structural stability correlated with the degree of Fc-glycan structure loss, especially at the CH2/CH3 domain interface. To improve analytical efficiency, rapid analytical methods are also favoured. Yang et al. (2016a) developed and compared two ultrafast methods for antibody glycan analysis that involved the rapid generation and purification of glycopeptides in either organic solvent or aqueous buffer followed by label-free quantification using MALDI-TOF MS. Yang et al. (2016b) combined high-resolution native MS and middle-down proteomics to analyze glycoprotein micro-heterogeneity by taking human erythropoietin (rhEPO) and properdin as model systems, which could bridge the gap between peptide- and protein-based MS platforms and contribute to the profiling of glycoproteins.

Moreover, it should be mentioned that establishing within-laboratory repeatability is critical to the harmonization of glycosylation analysis methods between-laboratories. To report and compare results for the full range of analytical methods presently used in the glycosylation analysis of mAbs, results of an interlaboratory study on the glycosylation of the primary sample of NISTmAb, a monoclonal antibody reference material, were reported (De Leoz et al. 2020). In this study, participation was open to all laboratories, regardless of experience or preferred analytical method. Protein glycosylation could be determined in various ways, including at the level of intact mAb, protein fragments, glycopeptides, or released glycans, using a wide variety of methods for derivatization, separation, identification, and quantification. Seventy-six laboratories from industry, university, research, government, and hospital sectors in Europe, North America, Asia, and Australia submitted a total of 103 reports on glycan analysis. The authors found that the results exhibited enormous diversity, e.g., the number of glycan compositions identified by each laboratory ranging from 4 to 48. A total of 116 glycan compositions were reported, in which 57 compositions could be assigned consensus abundance values. These consensus medians provided community-derived values for NIST-mAb primary sample. In addition, the better a laboratory's measurement precision, the more likely that the laboratory's mean values would agree with the community consensus. Measurement repeatability of more abundant glycans was relatively better, and the CVs increased with the decrease of glycan abundances. Reusch et al. (2015) described a thorough comparison of MS-based methods for glycan analysis using the same mAb sample, involving two laboratories, a biopharmaceutical company (Roche Diagnostics GmbH), and an academic research laboratory (Leiden University Medical Center). The mAb sample was analyzed 6-fold on two different days. In this study, 11 MS-based methods (7 methods using ESI ionization and 4 methods employing MALDI ionization) were evaluated for the analysis of the Fc glycosylation including precision, accuracy, throughput, and analysis time. Two methods detected glycosylated polypeptides after reduction or limited proteolytic cleavage at the IgG hinge region, six methods measured tryptic glycopeptides, and three methods analyzed PNGase F-released N-glycans. MS-based methods were compared with each other as well as with HILIC-UHPLC profiling of 2-AB-labelled glycans employing fluorescence detection, which served as a reference method. Special attention was paid to the measurement of low sialylation levels. They found that most methods showed excellent precision and accuracy, but some differences were also observed about the detection and quantitation of low abundant glycan species like the sialylated glycans and the amount of artefacts due to in-source decay.

6.4.2 Recombinant Protein Drugs

Stavenhagen et al. (2019) characterized site-specific N-glycosylation of atacicept including the glycosylation sites and their corresponding glycoforms using MS-based workflows, which is currently under clinical investigation for its biotherapeutic application in autoimmune diseases. They confirmed the presence of one N-glycosylation site, carrying 47 glycoforms covering 34 different compositions, next to 2 hinge region O-glycosylation sites with core 1-type glycans. Canis et al. (2018) studied the N-glycosylation of human plasma-derived factor VIII (pdFVIII) and six recombinant products (rFVIII) products expressed in CHO, BHK or HEK cell lines using a combination of MALDI MS and MS/MS, GC-MS and UPLC-UV-MS[E] technologies. They found that the entire N-glycan content of each sample appeared significantly different, although site-specific glycosylation of rFVIII were consistent with pdFVIII regardless of the expression system. Although the proportion of biologically important epitopes common to all samples (i.e., sialylation and high-mannose) varied between samples, some recombinant products expressed distinct and immunologically relevant epitopes, such as LacdiNAc, fucosylated LacdiNAc, NeuGc, Lewis X/Y, and $Gal_{\alpha1,3}Gal$ epitopes. And rFVIII expressed in HEK cells showed the greatest glycomic differences to human pdFVIII. Cowpera et al. (2018) compared 12 recombinant human rhEPO preparations from 11 manufacturers in China and 1 in Japan, in vivo biological activity, its relationship with glycosylation, and N-glycan mapping. They observed differences between glycosylation profiles (i.e., the varying occurrence of sialic acid O-acetylation, extension of N-glycan antennae with N-acetyllactosamine units) and the distribution of sialic acids across multi-antennary structures. And the presence of unusually high levels of suspected penta- and hexa-anionic N-glycans in several samples was consistent with elevated rhEPO isoform acidity and slightly elevated in vivo bioactivities. Zhu et al. (2017) performed site-specific characterization of the N-linked glycosylation on a set of human chorionic gonadotropin (hCG) drug products. Ten different lots and/or brands of commercial therapeutic hCGs were labelled with TMT 10plex reagents after tryptic digestion. The labelled intact glycopeptides were then analyzed by LC-MS with online alternating HCD/ETD/CID dissociation methods. Without considering other modifications, a total of 332 unique site-specific N-linked glycopeptides were identified. Most of the high-abundant glycans contained sialic acid. Their results showed that r-hCG and u-hCG products had distinct and conserved glycans, and 167 glycopeptides were significantly different between the naturally derived and recombinant hCG products. Thennati et al. (2018) also studied the comparative structural and biological attributes (with respect to its primary, higher-order structure, isoforms, charge variants, glycosylation, sialyation pattern, pharmacodynamics, and in vivo efficacy) of recombinant human chorionic gonadotropin, SB005 produced at Sun Pharmaceuticals, with reference product, Ovidrel® and Ovitrelle®. Stavenhagen et al. (2018) performed a site-specific N- and O-glycosylation analysis of plasma derived Human C1-inhibitor, which is a serine protease inhibitor and the major regulator of the contact activation pathway as well as the classical and lectin complement pathways. Cowper et al. (2020) characterized and compared the glycosylation profiles of five erythropoiesis stimulating agent products: Eprex® (epoetin alfa), NeoRecormon® (epoetin beta), Binocrit® (epoetin alfa biosimilar), Silapo (epoetin alfa biosimilar), and Aranesp® (darbepoetin alfa). They found that there were notable differences in N- and O-glycosylation, including attributes such as sialic acid occupation, O-acetylation, N-acetyllactosamine extended antennae, and sulphated/penta-sialylated N-glycans, which might cause divergence of therapeutic potencies. Seo et al. (2018) assessed the structural and functional similarity (e.g., glycan map) of Amgen's biosimilar ABP 215 and bevacizumab sourced from both the US and the EU, as well as the similarity between the US- and EU-sourced Bevacizumab. They found that ABP 215 was highly similar to bevacizumab with some minor differences in physicochemical attributes, and these differences did not affect functions relevant to the mechanism of action of ABP 215 and bevacizumab.

6.4.3 Vaccines

Comprehensive structural characterization of key virus proteins and innovative vaccine structural design followed by structural analysis are required to successfully develop efficient vaccines against corresponding infectious diseases. Thus, MS-based characterization is critical for both the vaccine design and quality control (Sharma et al. 2020).

6.4.3.1 Human Immunodeficiency Virus (HIV)

The HIV-1 envelope glycoprotein (Env) trimer consists of gp120 and gp41 subunits, among which gp120 is a major vaccine target, and knowing the site-specific glycosylation of gp120 can facilitate the rational design of glycopeptide antigens for HIV vaccine development. However, it is influenced by a variety of factors, including the genotype of the protein, the cell line used for expression, and the details of the construct design. Therefore, Panico et al. (2016) performed the glycosylation site analysis of gp120 derived from HIV-1 virions produced by infected T lymphoid cells. They observed that 20 of the 24 glycosylation sites in the gp120 were almost exclusively occupied with oligomannose glycans, two sites were a mixture of complex and hybrid glycans, one site carried a mixture of similar quantities of all three glycan classes, and one site was exclusively substituted with complex glycans. Go et al. (2017) mapped the glycosylation profile at every site in multiple HIV-1 Env trimmers and found that over half of the gp120 glycosylation sites on 11 different trimeric Envs had a conserved glycan profile, indicating that a native consensus glycosylation profile existed among trimers. And some soluble gp120s and gp140s exhibited highly divergent glycosylation profiles compared to trimeric Env. Wang et al. (2016b) characterized two HIV-1 subtype C gp120 Envs (1086.C and TV1.C) to confirm their sequence integrity, their biophysical immunogenicity, glycosylation patterns (both N-linked and O-linked), and disulfide linkages. Cao et al. (2017) carried out site specific N-glycosylation analysis of six different strains of HIV Env glycoprotein and found that most glycosites in recombinant Env trimers were fully occupied by glycans, varying in the proportion of high-mannose/hybrid and complex-type glycans. Behrens et al. (Behrens et al. 2016; Behrens et al. 2017) studied composition and antigenic effects of individual glycan sites of the trimeric HIV-1 envelope glycoprotein BG505 SOSIP.664. Struwe et al. (2017) analyzed N-glycan site occupancy of HIV-1 gp120 and found that gp120 monomers of the BG505 strain contained either fully occupied sequons or missing the equivalent of one and sometimes two glycans across the molecule.

Ion-mobility MS has emerged as a powerful method for structural characterization of released glycans because glycan ions are usually singly and doubly charged and have particular three-dimensional structures resulting in diverse ion mobility drift times (Bitto et al. 2015), which can be used to separate the isomeric N-linked glycan and profile complex type, high-mannose isomers in the trimeric envelop (Harvey et al. 2015). Therefore, Pritchard et al. (2015a) found that the glycosylation of native Env trimers isolated from peripheral blood mononuclear cells-derived virions consisted of mainly protein-directed oligomannose glycans with a population of highly processed sialylated bi-, tri-, and tetra-antennary glycans were capped with mostly α2,6-linked sialic acids. However, Env produced in HEK 293T cells failed to accurately reproduce the highly processed complex-type glycan structures, and in particular the precise linkage of sialic acid residues. In addition, they investigated the contribution of individual glycosylation sites in the formation of the so-called intrinsic mannose patch, and found that deletion of individual sites had a limited effect on the overall size of the intrinsic mannose patch but led to changes in the processing of neighbouring glycans (Pritchard et al. 2015b). These structural changes were largely tolerated by a panel of glycan-dependent broadly neutralizing antibodies targeting these regions, indicating a degree of plasticity in their recognition. Go et al. (2015) evaluated whether Env glycosylation was dependent on the Env form, i.e., membrane-anchored or soluble. They found that exogenous membrane-anchored Envs also displayed a virion-like glycan profile. Many of the sites contained exclusively high-mannose glycans, others retained complex glycans, resulting in a glycan profile that couldn't currently be mimicked on soluble gp120 or gp140 preparations.

6.4.3.2 Influenza

The influenza virus surface glycoprotein hemagglutinin (HA) is also a major target of host neutralizing antibodies. Parsons et al. (2016) characterized the glycosylation patterns of four recombinant H5 HAs derived from A/ Mallard/Denmark/64650/03 (H5N7) using MS-based methods. An et al. (2015b) examined HA N-glycosylation of H3N2 virus strains that they had engineered to closely mimic glycosylation sites gained between 1968 through 2002 starting with pandemic A/Hong Kong/1/68 (H3N2: HK68), including glycopeptide composition, sequence and site occupancy, and released glycans. She et al. (2017) evaluated the glycan composition, structural distribution, and topology of glycosylation for two high-yield candidate reassortant vaccines (NIBRG-121xp and NYMC-X181A) derived from the influenza virus strain A/California/7/2009. N-glycosylation can affect the host specificity, virulence, and infectivity of influenza A viruses. Chen et al. (2017b) proliferated one strain of the H9N2 subtypes in the embryonated chicken eggs (ECE) and human embryonic lung fibroblast cells (MRC-5) system. They found that the Fucα-1,6GlcNAc (core fucose) structure was increased, and penta-antennary N-glycans were only observed in the ECE system, while the sialylation α2,3/6-Gal structures were highly expressed and Fucα1,2Galβ1,4GlcNAc structures were only observed in the MRC-5 system. The existing sialylation α2,3/6 Gal sialoglycans made the offspring of the H9N2 virus preferentially attach to each other, which decreased viral virulence. Solano et al. (2017) quantified influenza neuraminidase activity after enzymatic cleavage of sialic acid from various substrates utilizing ultra-high performance liquid chromatography and isotope dilution MS.

6.4.3.3 Respiratory Syndrome Coronavirus 2 (SARS-CoV-2)

The emergence and rapid proliferation of the novel coronavirus, SARS-CoV-2, is a huge threat to global health, which results in urgent demand to find effective treatments including efficient and effective vaccine development.

The coronavirus spike (S) glycoprotein, which mediates viral attachment, entry, and membrane fusion, plays a critical role in the elicitation of the host immune response. It is comprised of two protein subunits (S1 and S2) and together possesses 22 potential N-glycosylation sites. To understand structure of the spike protein, Shajahan et al. (2020) profiled N-glycosylation on the spike protein subunits S1 and S2 expressed on human cells. Zhou et al. (2020b) analyzed recombinant SARS-CoV-2 spike protein secreted from BTI-Tn-5B1-4 insect cells. They acquired MS/ MS spectrums for glycopeptides of all 22 predicted N-glycosylated sites, which were all modified by high-mannose N-glycans. Watanabe et al. (2020) revealed the glycan structures on a recombinant SARS-CoV-2 S immunogen using a site-specific MS approach and indicated how SARS-CoV-2 S glycans differed from typical host glycan processing, which might have implications in viral pathobiology and vaccine design. Yang et al. (2020) used native MS to characterize ACE2/RBD complexes and evaluate the influence of heparin-related compounds (a synthetic pentasaccharide fondaparinux and a fixed-length eicosasaccharide heparin chains) on the stability of these complexes. Davies et al. (2020) compared the interactomes of non-structural protein-2 (nsp2) and nsp4 from three betacoronavirus strains: SARS-CoV-1, SARS-CoV-2, and hCoV-OC43-an endemic strain associated with the common cold, that were critical for virus replication. They identified common nsp2 interactors involved in endoplasmic reticulum (ER) Ca^{2+} signalling and mitochondria biogenesis; nsp4 interactors unique to each strain, such as E3 ubiquitin ligase complexes for SARS-CoV-1 and ER homeostasis factors for SARS-CoV-2; common nsp4 interactors included N-linked glycosylation machinery, unfolded protein response associated proteins, and anti-viral innate immune signalling factors. Both nsp2 and nsp4 interactors were strongly enriched in proteins localized at mitochondrial-associated ER membranes.

6.4.3.4 Others

Lei et al. (2015) characterized the structures of envelope protein N-linked glycans on mature Dengue viruses (DENV)-2 particles derived from insect cells, and found a high heterogeneity of DENV N-glycans, including mannose, GalNAc, GlcNAc, fucose, and sialic acid, among which high mannose-type and galactosylation were the major structures. Dubayle et al. (2015) compared the glycosylation pattern and the site-specific N-glycosylation of the E-protein from four serotypes of the Sanofi Pasteur tetravalent dengue vaccine (CYD) candidate using MS method. They found that the N-linked glycans of CYDs were a mix of high-mannose, hybrid, and complex glycans. High-mannose-type structures were predominant glycoforms at Asn67, while mainly complex- and hybrid-type structures existed at Asn153. Collar et al. (2017) analyzed five different strains of ebolavirus (BDBV, SUDV, TAFV, EBOV-Yambuku, and EBOV-Makona) produced in mammalian 293T cells and demonstrated that approximately 50 different N-glycan structures were present in envelope glycoprotein

($GP_{1,2}$) including high mannose, hybrid, and bi-, tri-, and tetra-antennary complex glycans with and without fucose and sialic acid. There was similar overall N-glycan composition between the different ebolavirus $GP_{1,2}$s. Lancaster et al. (2016) characterized the N-glycosylation patterns of Chikungunya virus (CHIKV)-like particles (VLPs), containing both E1 and E2 glycoproteins, derived from mammalian and insect cells. They found that the VLPs from HEK293 and SfBasic had significantly different N-glycosylation profiles. HEK293 derived CHIKV VLPs were modified by oligomannose, hybrid, and complex glycans, while VLPs derived from SfBasic were predominantly attached by oligomannose glycans. Zhao et al. (2017) characterized N-glycosylation of human enterovirus 71 virus-like particles expressed and purified from insect cells, which was a major causative pathogen of hand, foot, and mouth disease.

6.5 CONCLUSIONS AND PERSPECTIVES

Thanks to the development of sample preparation and analytical methods, sensitivity, accuracy, and robustness of protein glycosylation analysis has been greatly improved, contributing to both basic and clinical research. Of course, the applications of glycoproteomics and glycomics in clinical samples are far more than what we described here, there are many comprehensive reviews covering the importance of glycoproteomic and glycomic technologies in biomedical research (Ruhaak et al. 2018; Zhang et al. 2018). In terms of glycosylation analysis in clinical applications, extensive structural analysis (e.g., intact glycopeptides and glycoforms) will be one of important projects in the future for both biomarker discovery and evaluation of therapeutic glycoproteins. With the increasing application of N-glycoproteomics and N-glycomics in clinical research, it is essential to standardize the sample preparation and analytical procedure in order to obtain ■ confident results. Sample collection, sample preparation, and versatile analytical techniques suitable for trace samples are necessary so that we can make full use of precious samples to get reliable and comprehensive information. The automated and high-throughput workflow may be a good choice and helpful for leading to more robust results. N-glycoprotein and N-glycan analysis at a single-cell level is valuable to study cellular heterogeneity and diagnose diseases in the early stage. Furthermore, it is meaningful to integrate the existing data from clinical specimens, especially those data obtained from large sample sizes, and establish associated database.

REFERENCES

Aguilar, H. A., A. B. Iliuk, I. Chen, W. A. Tao. 2020. Sequential phosphoproteomics and N-glycoproteomics of plasma-derived extracellular vesicles. *Nature Protocol* 15(1): 161–80.

Al-wajeeh, A. S., M. N. Ismail, S. M. Salhimi, et al. 2017. Identification of glycobiomarker candidates for breast cancer using LTQ-orbitrap fusion technique. *International Journal of Pharmacology* 13: 425–37.

An, T. X., S. H. Qin, Y. Xu, et al. 2015a. Exosomes serve as tumour markers for personalized diagnostics owing to their important role in cancer metastasis. *Journal of Extracellular Vesicles* 4: 27522.

An, Y. M., J. A. McCullers, I. Alymova, L. M. Parsons, J. F. Cipollo. 2015b. Glycosylation analysis of engineered H3N2 influenza a virus hemagglutinins with sequentially added historically relevant glycosylation sites. *J Proteome Res* 14: 3957–69.

Bae, S., J. Brumbaugh, B. Bonavida. 2018. Exosomes derived from cancerous and non-cancerous cells regulate the anti-tumor response in the tumor microenvironment. *Genes Cancer* 9: 87–100.

Bai, H. H., Y. T. Pan, L. Qi, et al. 2018. Development a hydrazide-functionalized thermosensitive polymer based homogeneous system for highly efficient N-glycoprotein/glycopeptide enrichment from human plasma exosome. *Talanta* 186: 513–20.

Balmaña, M., A. Sarrats, E. Llop, et al. 2015. Identification of potential pancreatic cancer serum markers: increased sialyl-lewis X on ceruloplasmin. *Clinica Chimica Acta* 442: 56–62.

Bas, M., A. Terrier, E. Jacque, et al. 2019. Fc sialylation prolongs serum half-life of therapeutic antibodies. *J Immunol* 202(5): 1582–94.

Beck, A., S. Sanglier-Cianferani, A. Van Dorsselaer. 2012. Biosimilar, biobetter, and next generation antibody characterization by mass spectrometry. *Analytical Chemistry* 84(11): 4637–46.

Behrens, A., S. Vasiljevic, L. K. Pritchard, et al. 2016. Composition and antigenic effects of individual glycan sites of a trimeric HIV-1 envelope glycoprotein. *Cell Reports* 14(11): 2695–706.

Behrens, A., D. J. Harvey, E. Milne, et al. 2017. Molecular architecture of the cleavage-dependent mannose patch on a soluble HIV-1 envelope glycoprotein trimer. *Journal of Virology* 91: e01894-1616–..

Belczacka, I., M. Pejchinovski, M. Krochmal, et al. 2019. Urinary glycopeptide analysis for the investigation of novel biomarkers. *Proteomics: Clinical Applications* 13(3): 1800111.

Bellin, G., C. Gardin, L. Ferroni, et al. 2019. Exosome in cardiovascular diseases: a complex world full of hope. *Cells* 8(2): 166.

Biskup, K., E. I. Braicu, J. Sehouli, R. Tauber, V. Blanchard. 2017. The ascites N-glycome of epithelial ovarian cancer patients. *Journal of Proteomics* 157: 33–9.

Bitto, D., D. J. Harvey, S. Halldorsson, et al. 2015. Determination of N-linked glycosylation in viral glycoproteins by negative ion mass spectrometry and ion mobility. *Methods in Molecular Biology* 1331: 93–121.

Bladergroen, M. R., K. R. Reiding, A. L. H. Ederveen, et al. 2015. Automation of high-throughput mass spectrometry-based plasma N-glycome analysis with linkage-specific sialic acid esterification. *Journal of Proteome Research* 14: 4080–6.

Borelli, V., V. Vanhooren, E. Lonardi, et al. 2015. Plasma N-glycome signature of down syndrome. *Journal of Proteome Research* 14: 4232–45.

Canis, K., J. Anzengruber, E. Garenaux, et al. 2018. In-depth comparison of N-glycosylation of human plasma-derived factor VIII and different recombinant products: From structure to clinical implications. *Journal of Thrombosis and Haemostasis* 16(8): 1592–603.

Cao, W. Q., J. M. Huang, B. Y. Jiang, X. Gao, P. Y. Yang. 2016. Highly selective enrichment of glycopeptides based on zwitterionically functionalized soluble nanopolymers. *Scientific Reports* 6: 29776.

Cao, L. W., J. K. Diedrich, D. W. Kulp, et al. 2017. Global site-specific N-glycosylation analysis of HIV envelope glycoprotein. *Nature Communications* 8: 14954.

Cao, C. Y., L. Yu, D. M. Fu, J. L. Yuan, X. M. Liang. 2020. Absolute quantitation of high abundant Fc-glycopeptides from human serum IgG-1. *Analytica Chimica Acta* 1102: 130–9.

Caragata, M., A. K. Shah, B. L. Schulz, M. M. Hill, C. Punyadeera. 2016. Enrichment and identification of glycoproteins in human saliva using lectin magnetic bead arrays. *Analytical Biochemistry* 497: 76–82.

Chen, H. H., Z. A. Deng, C. C. Huang, H. M. Wu, X. Zhao, Y. Li. 2017a. Mass spectrometric profiling reveals association of N-glycan patterns with epithelial ovarian cancer progression. *Tumor Biology* 39(7): 1010428317716249.

Chen, W. T., Y. G. Zhong, R. Su, et al. 2017b. N-glycan profiles in H9N2 avian influenza viruses from chicken eggs and human embryonic lung fibroblast cells. *Journal of Virological Methods* 249: 10–20.

Chen, I., H. A. Aguilar, J. S. P. Paez, et al. 2018a. Analytical pipeline for discovery and verification of glycoproteins from plasma-derived extracellular vesicles as breast cancer biomarkers. *Analytical Chemistry* 90(10): 6307–13.

Chen, J. Y., H. H. Huang, S. Y. Yu, S. J. Wu, R. Kannagi, K. H. Khoo. 2018b. Concerted mass spectrometry-based glycomic approach for precision mapping of sulfo sialylated N-glycans on human peripheral blood mononuclear cells and lymphocytes. *Glycobiology* 28(1): 9–20.

Chen, H. F., C. Y. Shiao, M. Y. Wu, et al. 2020a. Quantitative determination of human IgA subclasses and their Fc-glycosylation patterns in plasma by using a peptide analogue internal standard and ultra-high-performance liquid chromatography/triple quadrupole mass spectrometry. *Rapid Communications in Mass Spectrometry* 34(S1): e8606.

Chen, S. Y., M. M. Dong, G. L. Yang, et al. 2020b. Glycans, glycosite, and intact glycopeptide analysis of N-linked glycoproteins using liquid handling systems. *Analytical Chemistry* 92: 1680–6.

Cheng, L., S. H. Gao, X. B. Song, et al. 2016. Comprehensive N-glycan profiles of hepatocellular carcinoma reveal association of fucosylation with tumor progression and regulation of FUT8 by microRNAs. *Oncotarget* 7: 61199–214.

Choi, N. Y., H. Hwang, E. S. Ji, et al. 2017. Direct analysis of site-specific N-glycopeptides of serological proteins in dried blood spot samples. *Analytical and Bioanalytical Chemistry* 409: 4971–81.

Clerc, F., M. Novokmet, V. Dotz, et al. 2018. Plasma N-glycan signatures are associated with features of inflammatory bowel diseases. *Gastroenterology* 155(3): 829–43.

Cocucci, E., J. Meldolesi. 2015. Ectosomes and exosomes: shedding the confusion between extracellular vesicles. *Trends in Cell Biology* 25: 364–72.

Collar, A. L., E. C. Clarke, E. Anaya, et al. 2017. Comparison of N- and O-linked glycosylation patterns of ebolavirus glycoproteins. *Virology* 502: 39–47.

Cotham, V. C., J. S. Brodbelt. 2016. Characterization of therapeutic monoclonal antibodies at the subunit-level using middle-down 193 nm ultraviolet photodissociation. *Analytical Chemistry* 88(7): 4004–13.

Cowper, B., M. Lavén, B. Hakkarainen, E. Mulugeta. 2020. Glycan analysis of erythropoiesis-stimulating agents. *Journal of Pharmaceutical and Biomedical Analysis* 180: 113031.

Cowpera, B., X. Li, L. Yu, Y. Zhou, W. H. Fan, C. M. Rao. 2018. Comprehensive glycan analysis of twelve recombinant human erythropoietin preparations from manufacturers in China and Japan. *Journal of Pharmaceutical and Biomedical Analysis* 153: 214–20.

Cymer, F., H. Beck, A. Rohde, D. Reusch. 2018. Therapeutic monoclonal antibody N-glycosylation - structure, function and therapeutic potential. *Biologicals* 52: 1–11.

Dalziel, M., M. Crispin, C. N. Scanlan, N. Zitzmann, R. A. Dwek. 2014. Emerging principles for the therapeutic exploitation of glycosylation. *Science* 343(6166): 1235681.

Darebna, P., P. Novak, R. Kucera, et al. 2017. Changes in the expression of N- and O-glycopeptides in patients with colorectal cancer and hepatocellular carcinoma quantified by full-MS scan FT-ICR and multiple reaction monitoring. *Journal of Proteomics* 153: 44–52.

Davies, J. P., K. M. Almasy, E. F. McDonald, L. Plate. 2020. Comparative multiplexed interactomics of SARS-CoV-2 and homologous coronavirus nonstructural proteins identifies unique and shared host-cell dependencies. *bioRxiv*. doi: 10.1101/2020.07.13.201517.

DeCoux, A., Y. Tian, K. Y. DeLeon-Pennell, et al. 2015. Plasma glycoproteomics reveals sepsis outcomes linked to distinct proteins in common pathways. *Critical Care Medicine* 43: 2049–58.

Dědová, T., E. I. Braicu, J. Sehouli, V. Blanchard. 2019. Sialic acid linkage analysis refines the diagnosis of ovarian cancer. *Frontiers in Oncology* 9: 261.

de-Freitas-Junior, J. C. M., J. Andrade-da-Costa, M. C. Silva, S. S. Pinho. 2017. Glycans as regulatory elements of the insulin/IGF system: Impact in cancer progression. *International Journal of Molecular Sciences* 18: 1921.

De Leoz, M. L. A, D. L. Duewer, A. Fung, et al. 2020. NIST interlaboratory study on glycosylation analysis of monoclonal antibodies: Comparison of results from diverse analytical methods. *Molecular & Cellular Proteomics* 19: 11–30.

de Vroome, S. W., S. Holst, M. R. Girondo, et al. 2018. Serum N-glycome alterations in colorectal cancer associate with survival. *Oncotarget* 9: 30610–23.

Dong, Q., X. J. Yan, Y. X. Liang, S. E. Stein. 2016. In-depth characterization and spectral library building of glycopeptides in the tryptic digest of a monoclonal antibody using 1D and 2D LC-MS/MS. *Journal of Proteome Research* 15: 1472–86.

Dong, X., S. Mondello, F. Kobeissy, F. Talih, R. Ferri, Y. Mechref. 2018. LC-MS/MS glycomics of idiopathic rapid eye movement sleep behavior disorder. *Electrophoresis* 39(24): 3096–103.

Dotz, V., R. F. H. Lemmers, K. R. Reiding, et al. 2018. Plasma protein N-glycan signatures of type 2 diabetes. *Biochim Biophys Acta Gen Subj* 1862: 2613–22.

Drake, R. R., C. McDowell, C. West, et al. 2020. Defining the human kidney N-glycome in normal and cancer tissues using MALDI imaging mass spectrometry. *Journal of Mass Spectrometry* 55: e4490.

Dubayle, J., S. Vialle, D. Schneider, et al. 2015. Site-specific characterization of envelope protein N-glycosylation on sanofi Pasteur's tetravalent CYD dengue vaccine. *Vaccine.* 33(11): 1360–8.

Elgundi, Z., M. Reslan, E. Cruz, V. Sifniotis, V. Kayser. 2017. The state-of-play and future of antibody therapeutics. *Advanced Drug Delivery Reviews* 122: 2–19.

Emmanouilidi, A., D. Paladin, D. W. Greening, M. Falasca. 2019. Oncogenic and non-malignant pancreatic exosome cargo reveal distinct expression of oncogenic and prognostic factors involved in tumor invasion and metastasis. *Proteomics* 19: e1800158.

Falck, D., B. C. Jansen, R. Plomp, D. Reusch, M. Haberger, M. Wuhrer. 2015. Glycoforms of immunoglobulin G based biopharmaceuticals are differentially cleaved by trypsin due to the glycoform influence on higher-order structure. *Journal of Proteome Research* 14(9): 4019–28.

Gao, Q., M. Dolikun, J. Štambuk, et al. 2017. Immunoglobulin G N-glycans as potential postgenomic biomarkers for hypertension in the Kazakh population. *Omics-a Journal of Integrative Biology* 21(7): 380–9.

Gbormittah, F. O., J. Bones, M. Hincapie, F. Tousi, W. S. Hancock, O. Iliopoulos. 2015. Clusterin glycopeptide variant characterization reveals significant site-specific glycan changes in the plasma of clear cell renal cell carcinoma. *Journal of Proteome Research* 14(6): 2425–36.

Gebrehiwot, A. G., D. S. Melka, Y. M. Kassaye, et al. 2018. Healthy human serum N-glycan profiling reveals the influence of ethnic variation on the identified cancer-relevant glycan biomarkers. *PLOS One* 13(12): e0209515.

Gebrehiwot, A. G., D. S. Melka, Y. M. Kassaye, et al. 2019. Exploring serum and immunoglobulin G Nglycome as diagnostic biomarkers for early detection of breast cancer in Ethiopian women. *BMC Cancer* 19: 588.

Giménez, E., M. Balmaña, J. Figueras, et al. 2015. Quantitative analysis of N-glycans from human alfa-acid-glycoprotein using stable isotope labeling and zwitterionic hydrophilic interaction capillary liquid chromatography electrospray mass spectrometry as tool for pancreatic disease diagnosis. *Analytica Chimica Acta* 866: 59–68.

Giorgetti, J., A. Beck, E. Leize-Wagner, Y. François. 2020. Combination of intact, middle-up and bottom-up levels to characterize 7 therapeutic monoclonal antibodies by capillary electrophoresis - mass spectrometry. *Journal of Pharmaceutical and Biomedical Analysis* 182: 113107.

Go, E. P., A. Herschhorn, C. Gu, et al. 2015. Comparative analysis of the glycosylation profiles of membrane-anchored HIV-1 envelope glycoprotein trimers and soluble gp140. *Journal of Virology* 89(16): 8245–57.

Go, E. P., H. T. Ding, S. J. Zhang, et al. 2017. Glycosylation benchmark profile for HIV-1 Envelope glycoprotein production based on eleven Env trimers. *Journal of Virology* 91(9): e02428–16.

Goyallon, A., S. Cholet, M. Chapelle, C. Junot, F. Fenaille. 2015. Evaluation of a combined glycomics and glycoproteomics approach for studying the major glycoproteins present in biofluids: Application to cerebrospinal fluid. *Rapid Commun. Mass Spectrom.* 29: 461–73.

Greabu, M., M. Battino, M. Mohora, et al. 2009. Saliva–a diagnostic window to the body, both in health and in disease. *Journal of Medicine and Life* 2: 124–32.

Groves, K., A. Cryar, S. Cowen, A. E. Ashcroft, M. Quaglia. 2020. Mass spectrometry characterization of higher order structural changes associated with the Fc-glycan structure of the NISTmAb reference material, RM 8761. *Journal of the American Society for Mass Spectrometry* 31: 553–64.

Guo, Z. G., X. J. Liu, M. L. Li, et al. 2015. Differential urinary glycoproteome analysis of type 2 diabetic nephropathy using 2D-LC-MS/MS and iTRAQ quantification. *Journal of Translational Medicine* 13(1): 371.

Guu, S. Y., T. H. Lin, S. C. Chang, et al. 2017. Serum N-glycome characterization and anti-carbohydrate antibody profiling in oral squamous cell carcinoma patients. *PLOS One* 12(6): e0178927.

Hafkenscheid, L., A. Bondt, H. U. Scherer, et al. 2017. Structural analysis of variable domain glycosylation of anti-citrullinated protein antibodies in rheumatoid arthritis reveals the presence of highly sialylated glycans. *Molecular & Cellular Proteomics* 16(2): 278–87.

Haga, Y., M. Uemura, S. Baba, et al. 2019. Identification of multisialylated LacdiNAc structures as highly prostate cancer specific glycan signatures on PSA. *Analytical Chemistry* 91(3): 2247–54.

Harper, L., B. C. Jacobs, C. O. S. Savage, R. Jefferis, A. M. Deelder, M. Morgan. 2015. Skewed Fc glycosylation profiles of anti-proteinase 3 immunoglobulin G1 autoantibodies from granulomatosis with polyangiitis patients show low levels of bisection, galactosylation, and sialylation. *Journal of Proteome Research* 14: 1657–65.

Harre, U., S. C. Lang, R. Pfeifle, et al. 2015. Glycosylation of immunoglobulin G determines osteoclast differentiation and bone loss. *Nature Communications* 6: 6651.

Harvey, D. J., M. Crispin, C. Bonomelli, J. H. Scrivens. 2015. Ion mobility mass spectrometry for ion recovery and clean-up of MS and MS/MS spectra obtained from low abundance viral samples. *Journal of the American Society for Mass Spectrometry* 26(10): 1754–67.

Higel, F., A. Seidl, F. Sörgel, W. Friess. 2016. N-glycosylation heterogeneity and the influence on structure, function and pharmacokinetics of monoclonal antibodies and Fc fusion proteins. *European Journal of Pharmaceutics and Biopharmaceutics* 100: 94–100.

Hinneburg, H., P. Korać, F. Schirmeister, et al. 2017. Unlocking cancer glycomes from histopathological formalin-fixed and paraffin-embedded (FFPE) tissue microdissections. *Molecular & Cellular Proteomics* 16(4): 524–36.

Ho, C. H., R. N. Chien, P. N. Cheng, et al. 2015. Aberrant serum immunoglobulin G glycosylation in chronic hepatitis B is associated with histological liver damage and reversible by antiviral therapy. *Journal of Infectious Diseases* 211: 115–24.

Holst, S., B. Heijs, N. de Haan, et al. 2016. Linkage- specific in situ sialic acid derivatization for N-glycan mass spectrometry imaging of formalin- fixed paraffin- embedded tissues. *Analytical Chemistry* 88: 5904–13.

Hong, Q., L. R. Ruhaak, C. Stroble, et al. 2015. A method for comprehensive glycosite-mapping and direct quantitation of serum glycoproteins. *Journal of Proteome Research* 14(12): 5179–92.

Hong, J., Y. Lee, C. Lee, et al. 2017. Physicochemical and biological characterization of SB2, a biosimilar of Remicade(R) (infliximab). *MAbs* 9(2): 364–82.

Hu, Y., C. R. Borges. 2017. A spin column-free approach to sodium hydroxide-based glycan permethylation. *Analyst* 142: 2748–59.

Hu, Y. W., P. Shah, D. J. Clark, M. H. Ao, H. Zhang. 2018. Reanalysis of global proteomic and phosphoproteomic data identified a large number of glycopeptides. *Analytical Chemistry* 90(13): 8065–71.

Huang, J., A. Guerrero, E. Parker, et al. 2015. Site-specific glycosylation of secretory immunoglobulin a from human colostrum. *Journal of Proteome Research* 14: 1335–49.

Huang, Y., Y. Nie, B. Boyes, R. Orlando. 2016. Resolving isomeric glycopeptide glycoforms with hydrophilic interaction chromatography (HILIC). *Journal of Biomolecular Techniques* 27: 98–104.

Huang, J., M. J. Kailemia, E. Goonatilleke, et al. 2017a. Quantitation of human milk proteins and their glycoforms using multiple reaction monitoring (MRM). *Analytical and Bioanalytical Chemistry* 409(2): 589–606.

Huang, Y., S. Zhou, J. Zhu, D. M. Lubman, Y. Mechref. 2017b. LC-MS/MS isomeric profiling of permethylated N-glycans derived from serum haptoglobin of hepatocellular carcinoma (HCC) and cirrhotic patients. *Electrophoresis* 38(17): 2160–7.

Huang, C. C., Y. M. Liu, H. M. Wu, D. H. Sun, Y. Li. 2017c. Characterization of IgG glycosylation in rheumatoid arthritis patients by MALDI-TOF-MS (n) and capillary electrophoresis. *Analytical and Bioanalytical Chemistry* 409(15): 3731–9.

Hurwitz, S. N., D. G. Meckes Jr. 2019. Extracellular vesicle integrins distinguish unique cancers. *Proteomes* 7: 14.

Jansen, B. C., A. Bondt, K. R. Reiding, et al. 2016. Pregnancy- associated serum N-glycome changes studied by high- throughput MALDI- TOF- MS. *Scientific Reports* 6: 23296.

Jia, X. W., J. Chen, S. S. Sun, et al. 2016. Detection of aggressive prostate cancer associated glycoproteins in urine using glycoproteomics and mass spectrometry. *Proteomics* 16: 2989–96.

Jia, G. Z., Z. Y. Dong, C. X. Sun, et al. 2017. Alterations in expressed prostate secretion-urine PSA N-glycosylation discriminate prostate cancer from benign prostate hyperplasia. *Oncotarget* 8(44): 76987–99.

Jia, L. Y., J. Zhang, T. R. Ma, Y. Y. Guo, Y. Yu, J. H. Cui. 2018. The function of fucosylation in progression of lung cancer. *Frontiers in Oncology* 8: 565.

Jiang, B., Q. Wu, N. Deng, et al. 2016a. Hydrophilic GO/Fe$_3$O$_4$/Au/PEG nano-composites for highly selective enrichment of glycopeptides. *Nanoscale* 8: 4894–7.

Jiang, H., H. M. Yuan, Y. Y. Qu, et al. 2016b. Preparation of hydrophilic mono-lithic capillary column by in situ photo-polymerization of N-vinyl-2-pyrrolidinone and acrylamide for highly selective and sensitive enrichment of N-linked glycopeptides. *Talanta* 146: 225–30.

Jie, J. Z., D. Liu, B. Yang, X. J. Zou. 2020. Highly efficient enrichment method for human plasma glycoproteome analyses using tandem hydrophilic interaction liquid chromatography workflow. *Journal of Chromatography A* 1610: 460546.

Jóźwik, J., J. Kałużna-Czaplińska. 2016. Current applications of chromatographic methods in the study of human body fluids for diagnosing disorders. *Critical Reviews in Analytical Chemistry* 46(1): 1–14.

Kailemia, M. J., D. Park, C. B. Lebrilla. 2017. Glycans and glycoproteins as specific biomarkers for cancer. *Analytical and Bioanalytical Chemistry* 409(2): 395–410.

Kammeijer, G. S. M., J. Nouta, J. J. M. C. H. de la Rosette, T. M. de Reijke, M. Wuhrer. 2018. An in-depth glycosylation assay for urinary prostate-specific antigen. *Analytical Chemistry* 90(7): 4414–21.

Kang, H., T. J. Tolbert, C. Schöneich. 2019. Photoinduced tyrosine side chain fragmentation in IgG4-Fc: mechanisms and solvent isotope effects. *Molecular Pharmaceutics* 16(1): 258–72.

Kang, H., N. R. Larson, D. R. White, C. R. Middaugh, T. Tolbert, C. Schöneich. 2020. Effects of Glycan Structure on the stability and receptor binding of an IgG4-Fc. *Journal of Pharmaceutical Sciences* 109: 677–89.

Kawaguchi-Sakita, N., K. Kaneshiro-Nakagawa, M. Kawashima, et al. 2016. Serum immunoglobulin G fc region N-glycosylation profiling by matrix-assisted laser desorption/ionization mass spectrometry can distinguish breast cancer patients from cancer-free controls. *Biochemical and Biophysical Research Communications* 469(4): 1140–5.

Kawahara, R., F. Ortega, L. Rosa-Fernandes, et al. 2018. Distinct urinary glycoprotein signatures in prostate cancer patients. *Oncotarget* 9(69): 33077–97.

Kazuno, S., J. Furukawa, Y. Shinohara, et al. 2016. Glycosylation status of serum immunoglobulin G in patients with prostate diseases. *Cancer medicine* 5(6): 1137–46.

Kim, S., J. S. Song, S. Park, et al. 2017a. Drifts in ADCC-related quality attributes of Herceptin(R): impact on development of a trastuzumab biosimilar. *MAbs* 9(4): 704–14.

Kim, J., S. H. Lee, S. Choi, et al. 2017b. Direct analysis of aberrant glycosylation on haptoglobin in patients with gastric cancer. *Oncotarget* 8(7): 11094–104.

Kim, J., C. W. Park, D. Um, et al. 2017c. Mass spectrometric screening of ovarian cancer with serum glycans. *Disease Markers* 2014: 634289.

Kim, Y. M., J. S. Park, S. K. Kim, et al. 2018. The transgenic chicken derived anti-CD20 monoclonal antibodies exhibits greater anti-cancer therapeutic potential with enhanced Fc effector functions. *Biomaterials* 167: 58–68.

Kim, K. H., G. W. Park, J. E. Jeong, et al. 2019. Parallel reaction monitoring with multiplex immunoprecipitation of N-glycoproteins in human serum for detection of hepatocellular carcinoma. *Analytical and Bioanalytical Chemistry* 411(14): 3009–19.

Kirwan, A., M. Utratna, M. E. O'Dwyer, L. Joshi, M. Kilcoyne. 2015. Glycosylation-based serum biomarkers for cancer diagnostics and prognostics. *Biomed Research International* 2015: 490531.

Lamanna, W. C., J. Holzmann, H. P. Cohen, et al. 2018. Maintaining consistent quality and clinical performance of biopharmaceuticals. *Current Opinion in Chemical Biology* 18(4): 369–79.

Lancaster, C., P. Pristatsky, V. M. Hoang, et al. 2016. Characterization of N-glycosylation profiles from mammalian and insect cell derived chikungunya VLP. *Journal of Chromatography B* 1032: 218–23.

Lee, J., S. Hua, S. H. Lee, et al. 2018a. Designation of fingerprint glycopeptides for targeted glycoproteomic analysis of serum haptoglobin: Insights into gastric cancer biomarker discovery. *Analytical and Bioanalytical Chemistry* 410: 1617–29.

Lee, K. H., J. Lee, J. S. Bae, et al. 2018b. Analytical similarity assessment of rituximab biosimilar CT-P10 to reference medicinal product. *MAbs* 10(3):380–96.

Lei, Y., H. Yu, Y. Dong, et al. 2015. Characterization of N-glycan structures on the surface of mature dengue 2 virus derived from insect cells. *PLOS One* 10(7): e0132122.

Leijon, H., T. Kaprio, A. Heiskanen, et al. 2017. N-glycomic profiling of pheochromocytomas and paragangliomas separates metastatic and nonmetastatic disease. *The Journal of clinical endocrinology and metabolism* 102(11): 3990–4000.

Li, H. H., W. J. Gao, X. J. Feng, B. F. Liu, X. Liu. 2016. MALDI-MS analysis of sialylated N-glycan linkage isomers using solid-phase two step derivatization method *Analytical Chimica Acta* 924: 77–85.

Li, Q. K., P. Shah, Y. Tian, et al. 2017. An integrated proteomic and glycoproteomic approach uncovers differences in glycosylation occupancy from benign and malignant epithelial ovarian tumors. *Clinical Proteomics* 14: 16.

Li, D. J., H. J. Xia, L. Wang. 2018. Branched polyethyleneimine-assisted boronic acid-functionalized silica nanoparticles for the selective enrichment of trace glycoproteins. *Talanta* 184: 235–43.

Li, F., J. Ding. 2019. Sialylation is involved in cell fate decision during development, reprogramming and cancer progression. *Protein Cell* 10(8): 550–65.

Li, X., J. J. Xu, M. T. Li, X. F. Zeng, J. Wang, C. J. Hu. 2019. Aberrant glycosylation in autoimmune disease. *Clinical and Experimental Rheumatology* 38(4): 767–75.

Liang, Y., C. Wu, Q. Zhao, et al. 2015. Gold nanoparticles immobilized hydrophilic monoliths with variable functional modification for highly selective enrichment and on-line deglycosylation of glycopeptides. *Analytica Chimica Acta* 900: 83–9.

Lin, C. W., M. H. Tsai, S. T. Li, et al. 2015. A common glycan structure on immunoglobulin G for enhancement of effector functions. *Proceedings of the National Academy of Sciences USA* 112: 10611–6.

Liu, L. 2015. Antibody glycosylation and its impact on the pharmacokinetics and pharmacodynamics of monoclonal antibodies and Fc fusion proteins. *Journal of Pharmaceutical Sciences* 104(6): 1866–84.

Liu, T. H., D. H. Liu, R. Q. Liu, et al. 2017a. Discovering potential serological biomarker for chronic hepatitis B virus-related hepatocellular carcinoma in Chinese population by MAL-associated serum glycoproteomics analysis. *Scientific Reports* 7: 38918.

Liu, T. H., S. X. Shang, W. Li, et al. 2017b. Assessment of hepatocellular carcinoma metastasis glycobiomarkers using advanced quantitative N-glycoproteome analysis. *Frontiers in Physiology* 8: 472.

Liu, S., W. J. Gao, Y. Wang, et al. 2017c. Comprehensive N-glycan profiling of cetuximab biosimilar candidate by NP-HPLC and MALDI-MS. *PLoS One* 12(1): e0170013.

Liu, Y. F., C. Wang, R. Wang et al. 2018a. Isomer-specific profiling of N-glycans derived from human serum for potential biomarker discovery in pancreatic cancer. *Journal of Proteomics* 181: 160–9.

Liu, J. N., M. Dolikun, J. Štambuk, et al. 2018b. The association between subclass-specific IgG Fc N-glycosylation profiles and hypertension in the Uygur, Kazak, Kirgiz, and Tajik populations. *Journal of Human Hypertension* 32: 555–63.

Liu, S., L. M. Cheng, Y. Fu, B. F. Liu, X. Liu. 2018c. Characterization of IgG N-glycome profile in colorectal cancer progression by MALDI-TOF-MS. *Journal of Proteomics* 181: 225–37.

Liu, S., Z. W. Huang, Q. W. Zhang, et al. 2020. Profiling of isomer-specific IgG N-glycosylation in cohort of Chinese colorectal cancer patients. *BBA - General Subjects* 1864: 129510.

Losev, Y., A. Paul, M. Frenkel-Pinter, et al. 2019. Novel model of secreted human tau protein reveals the impact of the abnormal N-glycosylation of tau on its aggregation propensity. *Scientific Reports* 9: 2254.

Ma, W., L. N. Xu, Z. Li, Y. L. Sun, Y. Bai, H. W. Liu. 2016. Post-synthetic modification of an amino-functionalized metal-organic framework for highly efficient enrichment of N-linked glycopeptides. *Nanoscale* 8: 10908.

Ma, J. F., M. Sanda, R. H. Z. Wei, L. H. Zhang, R. Goldman. 2018. Quantitative analysis of core fucosylation of serum proteins in liver diseases by LC-MS-MRM. *Journal of Proteomics* 189: 67–74.

Maier, M., D. Reusch, C. Bruggink, P. Bulau, M. Wuhrer, M. Mølhøj. 2016. Applying mini-bore HPAEC-MS/MS for the characterization and quantification of Fc N-glycans from heterogeneously glycosylated IgGs. *J Chromatography B Analytical Technologies in the Biomedical and Life Science* 1033-1034: 342–52.

Martín, S. M., C. Delporte, A. Farrell, N. N. Iglesias, N. McLoughlin, J. Bones. 2015. Comparative analysis of monoclonal antibody N-glycosylation using stable isotope labelling and UPLC-fluorescence-MS. *Analyst* 140: 1442–7.

Melo, S. A., L. B. Luecke, C. Kahlert, et al. 2015. Glypican-1 identifies cancer exosomes and detects early pancreatic cancer. *Nature* 523: 177–82.

Miyamoto, S., C. D. Stroble, S. Taylor, et al. 2018. Multiple reaction monitoring for the quantitation of serum protein glycosylation profiles: Application to ovarian cancer. *Journal of Proteome Research* 17: 222–33.

Möginger, U., S. Grunewald, R. Hennig, et al. 2018. Alterations of the human skin N- and O-glycome in basal cell carcinoma and squamous cell carcinoma. *Frontiers in Oncology* 8: 70.

More, A. S., R. T. Toth, S. Z. Okbazghi, et al. 2018. Impact of glycosylation on the local backbone flexibility of well-defined IgG1-Fc glycoforms using hydrogen exchange-mass spectrometry. *Journal of Pharmaceutical Sciences* 107(9): 2315–24.

Mozziconacci, O., S. Okbazghi, A. S. More, D. B. Volkin, T. Tolbert, C. Sch€oneich. 2016. Comparative evaluation of the chemical stability of 4 well-defined immunoglobulin G1-Fc glycoforms. *Journal of Pharmaceutical Sciences* 105(2): 575–87.

Munkley, J., D. J. Elliott. 2016. Hallmarks of glycosylation in cancer. *Oncotarget* 7(23): 35478–89.

Narita, T., S. Hatakeyama, T. Yoneyama, et al. 2017. Clinical implications of serum N-glycan profiling as a diagnostic and prognostic biomarker in germ-cell tumors. *Cancer Medicine* 6(4): 739–48.

Nigjeh, E. N., R. Chen, Y. Allen-Tamura, R. E. Brand, T. A. Brentnall, S. Pan. 2017. Spectral library-based glycopeptide analysis-detection of circulating galectin-3 binding protein in pancreatic cancer. *Proteomics Clinical Applications* 11: 9–10.

Noro, D., T. Yoneyama, S. Hatakeyama, et al. 2017. Serum aberrant N-glycan profile as a marker associated with early antibody-mediated rejection in patients receiving a living donor kidney transplant. *International Journal of Molecular Sciences* 18(8): 1731.

Pabst, M., S. K. Küster, F. Wahl, J. Krismer, P. S. Dittrich, R. Zenobi. 2015. A microarray-matrix-assisted laser desorption/ionization mass spectrometry approach for site-specific protein N-glycosylation analysis, as demonstrated for human serum immunoglobulin M (IgM). *Molecular & Cellular Proteomics* 14: 1645–56.

Palmigiano, A., R. Barone, L. Sturiale, et al. 2016. CSF N-glycoproteomics for early diagnosis in Alzheimer's disease. *Journal of Proteomics* 131: 29–37.

Pan, S., T. A. Brentnall, R. Chen. 2016. Glycoproteins and glycoproteomics in pancreatic cancer. *World Journal of Gastroenterology* 22(42): 9288–99.

Panico, M., L. Bouché, D. Binet, et al. 2016. Mapping the complete glycoproteome of virion-derived HIV-1 gp120 provides insights into broadly neutralizing antibody binding. *Scientific Reports* 6: 32956.

Parsons, L. M., Y. M. An, R. P. de Vries, C. A. M. de Haan, J. F. Cipollo. 2016. Glycosylation characterization of an influenza H5N7 hemagglutinin series with engineered glycosylation patterns: Implications for structure-function relationships. *Journal of Proteome Research* 16: 398–412.

Peng, Y., L. M. Wang, Y. Zhang, H. M. Bao, H. J. Lu. 2019. Stable isotope sequential derivatization for linkage specific analysis of sialylated N-glycan isomers by MS. *Analytical Chemistry* 91(24): 15993–6001.

Planinc, A., B. Dejaegher, Y. V. Heyden, et al. 2017. Batch-to-batch N-glycosylation study of infliximab, trastuzumab and bevacizumab, and stability study of bevacizumab. *European Journal of Hospital Pharmacy-Science and Practice* 24(5): 286–92.

Plomp, R., L. R. Ruhaak, H. W. Uh, et al. 2017. Subclass-specific IgG glycosylation is associated with markers of inflammation and metabolic health. *Scientific Reports* 7: 12325.

Plomp, R., N. de Haan, A. Bondt, J. Murli, V. Dotz, M. Wuhrer. 2018. Comparative glycomics of immunoglobulin A and G from saliva and plasma reveals biomarker potential. *Frontiers in Immunology* 9: 2436.

Powers, T. W., S. Holst, M. Wuhrer, A. S. Mehta, R. R. Drake. 2015. Two-dimensional N-glycan distribution mapping of hepatocellular carcinoma tissues by MALDI-imaging mass spectrometry. *Biomolecules* 5: 2554–72.

Pritchard, L. K., D. J. Harvey, C. Bonomelli, M. Crispin, K. J. Doores. 2015a. Cell-and protein-directed glycosylation of native cleaved HIV-1 envelope. *Journal of Virology* 89: 8932–44.

Pritchard, L. K., D. I. R. Spencer, L. Royle, et al. 2015b. Glycan clustering stabilizes the mannose patch of HIV-1 and preserves vulnerability to broadly neutralizing antibodies. *Nature Communications* 6: 7479.

Qin, Y. N., Y. G. Zhong, T. R. Ma, et al. 2017a. A pilot study of salivary N-glycome in HBV-induced chronic hepatitis, cirrhosis, and hepatocellular carcinoma. *Glycoconjugate Journal* 34(4): 523–35.

Qin, R. H., J. J. Zhao, W. J. Qin, et al. 2017b. Discovery of non-invasive glycan biomarkers for detection and surveillance of gastric cancer. *Journal of Cancer* 8(10): 1908–16.

Qin, Y. N., Y. N. Chen, J. Yang, et al. 2017c. Serum glycopattern and *Maackia amurensis* lectin-II binding glycoproteins in autism spectrum disorder. *Scientific Reports* 7: 46041.

Qin, H. Q., Y. Chen, J. W. Mao, et al. 2019. Proteomics analysis of site-specific glycoforms by a virtual multistage mass spectrometry method. *Analytical Chimica Acta* 1070: 60–8.

Reiding, K. R., L. R. Ruhaak, H. W. Uh, et al. 2017. Human plasma N-glycosylation as analyzed by matrix-assisted laser desorption/ionization-Fourier transform ion cyclotron resonance-MS associates with markers of inflammation and metabolic health. *Molecular & Cellular Proteomics* 16(2): 228–42.

Reiding, K. R., G. C. M. Vreeker, A. Bondt, et al. 2018. Serum protein N-glycosylation changes with rheumatoid arthritis disease activity during and after pregnancy. *Frontiers of Medicine* 4: 241.

Reiding, K. R., X. V. Franc, M. G. Huitema, E. Brouwer, P. Heeringa, A. J. R. Heck. 2019a. Neutrophil myeloperoxidase harbors distinct site-specific peculiarities in its glycosylation. *The Journal of Biological Chemistry* 294(52): 20233–45.

Reiding, K. R., A. Bondt, R. Hennig, et al. 2019b. High- throughput serum N-glycomics: method comparison and application to study rheumatoid arthritis and pregnancy- associated changes. *Molecular & Cellular Proteomics* 18: 3–15.

Reusch, D., M. Haberger, D. Falck, et al. 2015. Comparison of methods for the analysis of therapeutic immunoglobulin G Fc-glycosylation profiles-part 2: Mass spectrometric methods. *MAbs* 7: 732–42.

Rodrigues, J. G., M. Balmaña, J. A. Macedo, et al. 2018. Glycosylation in cancer: Selected roles in tumour progression, immune modulation and metastasis. *Cellular Immunology* 333: 46–57.

Rombouts, Y., A. Willemze, J. J. B. C. van Beers, et al. 2015a. Extensive glycosylation of ACPA-IgG variable domains modulates binding to citrullinated antigens in rheumatoid arthritis. *Annals of the Rheumatic Diseases* 75: 578–85.

Rombouts, Y., E. Ewing, L. A. van de Stadt, et al. 2015b. Anti-citrullinated protein antibodies acquire a pro-inflammatory Fc glycosylation phenotype prior to the onset of rheumatoid arthritis. *Annals of the Rheumatic Diseases* 74(1): 234–41.

Rouwendal, G. J., M. M. van der Lee, S. Meyer, et al. 2016. A comparison of anti-HER2 IgA and IgG1 in vivo efficacy is facilitated by high N-glycan sialylation of the IgA. *MAbs* 8: 74–86.

Ruhaak, L. R., D. A. Barkauskas, J. Torres, et al. 2015. The serum immunoglobulin G glycosylation signature of gastric cancer. *EuPA Open Proteomics* 6: 1–9.

Ruhaak, L. R., C. Stroble, J. L. Dai, et al. 2016a. Serum glycans as risk markers for non-small cell lung cancer. *Cancer Prevention Research* 9(4): 317–23.

Ruhaak, L. R., K. Kim, C. Stroble, et al. 2016b. Protein-specific differential glycosylation of immunoglobulins in serum of ovarian cancer patients. *Journal of Proteome Research* 15(3): 1002–10.

Ruhaak, R. L., G. Xu, Q. Li, E. Goonatilleke, C. B. Lebrilla. 2018. Mass spectrometry approaches to glycomic and glycoproteomic analyses. *Chemical Reviews* 118: 7886–930.

Russell, A. C., M. Simurina, M. T. Garcia, et al. 2017. The N-glycosylation of immunoglobulin G as a novel biomarker of Parkinson's disease. *Glycobiology* 27(5): 501–10.

Saarinen, L., P. Nummela, H. Leinonen, et al. 2018. Glycomic profiling highlights increased fucosylation in pseudomyxoma peritonei. *Molecular & Cellular Proteomics* 17(11): 2107–18.

Sajic, T., Y. Liu, E. Arvaniti, et al. 2018. Similarities and differences of blood N-glycoproteins in five solid carcinomas at localized clinical stage analyzed by SWATH-MS. *Cell Reports* 23(9): 2819–31.

Sajid, M. S., B. Jovcevski, T. L. Pukala, F. Jabeen, M. Najam-ul-Haq. 2020. Fabrication of piperazine functionalized polymeric monolithic tip for rapid enrichment of glycopeptides/glycans. *Analytical Chemistry* 92: 683–9.

Saldova, R., H. Stöckmann, R. O'Flaherty, D. J. Lefeber, J. Jaeken, P. M. Rudd. 2015. N-glycosylation of serum IgG and total glycoproteins in MAN1B1 deficiency. *Journal of Proteome Research* 14: 4402–12.

Schauer, R. 2009. Sialic acids as regulators of molecular and cellular interactions. *Current Opinion in Structural Biology* 19: 507–14.

Schedin-Weiss, S., S. Gaunitz, P. Sui, et al. 2020. Glycan biomarkers for Alzheimer disease correlate with T-tau and P-tau in cerebrospinal fluid in subjective cognitive impairment. *The FEBS Journal* 287(15): 3221–34.

Scott, D. A., K. Norris-Caneda, L. Spruill, et al. 2019. Specific N-linked glycosylation patterns in areas of necrosis in tumor tissues. *International Journal of Mass Spectrometry* 437: 69–76.

Seo, N., A. Polozova, M. X. Zhang, et al. 2018. Analytical and functional similarity of Amgen biosimilar ABP 215 to bevacizumab. *MAbs* 10(4): 678–91.

Sethi, M. K., H. Kim, C. K. Park, et al. 2015. In-depth N-glycome profiling of paired colorectal cancer and non-tumorigenic tissues reveals cancer-, stage- and EGFR-specific protein N-glycosylation. *Glycobiology* 25(10): 1064–78.

Shah, A. K., K. A. LêCao, E. Choi, et al. 2015. Serum glycoprotein biomarker discovery and qualification pipeline reveals novel diagnostic biomarker candidates for esophageal adenocarcinoma. *Molecular & Cellular Proteomics* 14: 3023–39.

Shaha, I. S., S. Lovell, N. Mehzabeen, K. P. Battaile, T. J. Tolbert. 2017. Structural characterization of the Man5 glycoform of human IgG3 Fc *Molecular Immunology* 92: 28–37.

Shajahan, A., N. T. Supekar, A. S. Gleinich, P. Azadi. 2020. Deducing the N- and O- glycosylation profile of the spike protein of novel coronavirus SARS-CoV-2. *Glycobiology* doi: 10.1093/glycob/cwaa042.

Shan, X., W. J. Dong, L. Zhang, et al. 2020. Role of fucosyltransferase IV in the migration and invasion of human melanoma cells. *IUBMB Life* 72: 942–56.

Sharma, V. K., I. Sharma, J. Glick. 2020. The expanding role of mass spectrometry in the field of vaccine development. *Mass Spectrometry Reviews* 39: 83–104.

She, Y. M., A. Farnsworth, X. G. Li, T. D. Cyr. 2017. Topological N-glycosylation and site-specific N-glycan sulfation of influenza proteins in the highly expressed H1N1 candidate vaccines. *Scientific Reports* 7: 10232.

Shi, Z. H., L. Y. Pu, Y. S. Guo, et al. 2017. Boronic acid-modified magnetic $Fe_3O_4@$ $mTiO_2$ microspheres for highly sensitive and selective enrichment of N-glycopeptides in amniotic fluid. *Scientific Reports* 7: 4603.

Silsirivanit, A. 2019. Glycosylation markers in cancer. *Advances in Clinical Chemistry* 89: 189–213.

Šimurina, M., N. de Haan, F. Vučković, et al. 2018. Glycosylation of immunoglobulin G associates with clinical features of inflammatory bowel diseases. *Gastroenterology* 154(5): 1320–33.

Singh, S., N. K. Kumar, P. Dwiwedi, et al. 2018. Monoclonal antibodies: A review. *Current Clinical Pharmacology* 13(2): 85–99.

Solakyildirim, K., L. Y. Li, R. J. Linhardt. 2018. Mass spectrometry in the determination of glycosylation site and N-glycan structures of human placental alkaline phosphatase. *Mass Spectrometry Letters* 9(3): 67–72.

Solano, M. I., A. R. Woolfitt, T. L. Williams, et al. 2017. Quantification of influenza neuraminidase activity by ultra-high performance liquid chromatography and isotope dilution mass spectrometry. *Analytical Chemistry* 89: 3130–7.

Song, T., D. Aldredge, C. B. Lebrilla. 2015. A method for in-depth structural annotation of human serum glycans that yields biological variations. *Analytical Chemistry* 87: 7754–62.

Song, T., P. Chen, C. Stroble, et al. 2018. Serum glycosylation characterization of osteonecrosis of the femoral head by mass spectrometry. *European Journal of Mass Spectrometry* 24: 178–87.

Sonneveld, M. E., M. de Haas, C. Koeleman, et al. 2017. Patients with IgG1-anti-red blood cell autoantibodies show aberrant Fc glycosylation. *Scientific Reports* 7(1): 8187.

Spiciarich, D. R., R. Nolley, S. L. Maund, et al. 2017. Bioorthogonal labeling of human prostate cancer tissue slice cultures for glycoproteomics. *Angewandte Chemie International Edition* 56(31): 8992–7.

Stavenhagen, K., R. Plomp, M. Wuhrer. 2015. Site-specific protein N- and O-glycosylation analysis by a C18-porous graphitized carbon liquid chromatography-electrospray ionization mass spectrometry approach using pronase treated glycopeptides. *Analytical Chemistry* 87: 11691–9.

Stavenhagen, K., H. M. Kayili, S. Holst, et al. 2018. N- and O-glycosylation analysis of human C1-inhibitor reveals extensive mucin-type O-glycosylation. *Molecular & Cellular Proteomics* 17(6): 1225–38.

Stavenhagen, K., R. Gahoual, E. D. Vega, et al. 2019. Site-specific N- and O-glycosylation analysis of atacicept. *MAbs* 11(6): 1053–63.

Struwe, W. B., A. Stuckmann, A. J. Behrens, K. Pagel, M. Crispin. 2017. Global N-glycan site occupancy of HIV-1 gp120 by metabolic engineering and high-resolution intact mass spectrometry. *ACS Chemical Biology* 12(2): 357–61.

Sun, N. R., J. W. Wang, J. Z. Yao, C. H. Deng. 2017. Hydrophilic mesoporous silica materials for highly specific enrichment of N-linked glycopeptide. *Analytical Chemistry* 89: 1764–71.

Sun, S. S., Y. W. Hu, L. Jia, et al. 2018. Site-specific profiling of serum glycoproteins using N-linked glycan and glycosite analysis revealing atypical N-glycosylation sites on albumin and alpha-1B-glycoprotein. *Analytical Chemistry* 90(10): 6292–9.

Takahashi, S., T. Sugiyama, M. Shimomura, et al. 2016. Site-specific and linkage analyses of fucosylated N-glycans on haptoglobin in sera of patients with various types of cancer: Possible implication for the differential diagnosis of cancer. *Glycoconjugate Journal* 33: 471–82.

Talabnin, K., C. Talabnin, M. Ishihara, P. Azadi. 2018. Increased expression of the high-mannose M6N2 and NeuAc3H3N3M3N2F tri-antennary N-glycans in cholangiocarcinoma. *Oncology Letters* 15(1): 1030–6.

Tan, Z., H. Yin, S. Nie, et al. 2015. Large-scale identification of core-fucosylated glycopeptide sites in pancreatic cancer serum using mass spectrometry. *Journal of Proteome Research* 14(4): 1968–78.

Tanabe, K., K. Kitagawa, N. Kojima, S. Iijima. 2016. Multi-fucosylated alpha-1-acid glycoprotein as a novel marker for hepatocellular carcinoma. *Journal of Proteome Research* 15: 2935–44.

Thennati, R., S. K. Singh, N. Nage, et al. 2018. Analytical characterization of recombinant hCG and comparative studies with reference product. *Biologics: Targets and Therapy* 12: 23–35.

Theodoratou, E., K. Thaçi, F. Agakov, et al. 2016. Glycosylation of plasma IgG in colorectal cancer prognosis. *Scientific Reports* 6: 28098.

Togayachi, A., J. Iwaki, H. Kaji, et al. 2017. Glycobiomarker, fucosylated short-form secretogranin III levels are increased in serum of patients with small cell lung carcinoma. *Journal of Proteome Research* 16(12): 4495–505.

Tran, B. Q., C. Barton, J. H. Feng, et al. 2015. Comprehensive glycosylation profiling of IgG and IgG-fusion proteins by top-down MS with multiple fragmentation techniques. *Journal of Proteomics* 134: 93–101.

Vajaria, B. N., P. S. Patel. 2017. Glycosylation: A hallmark of cancer? *Glycoconjugate Journal* 34: 147–56.

van der Burgt, Y. E. M., D. P. A. Kilgour, Y. O. Tsybin, et al. 2019. Structural analysis of monoclonal antibodies by ultrahigh resolution MALDI in-source decay FT-ICR mass spectrometry. *Analytical Chemistry* 91(3): 2079–85.

van Scherpenzeel, M., G. Steenbergen, E. Morava, R. A. Wevers, D. J. Lefeber. 2015. High-resolution mass spectrometry glycoprofiling of intact transferrin for diagnosis and subtype identification in the congenital disorders of glycosylation. *Translational Research* 166: 639–49.

Váradi, C., C. Jakes, J. Bones. 2020. Analysis of cetuximab N-glycosylation using multiple fractionation methods and capillary electrophoresis mass spectrometry. *Journal of Pharmaceutical and Biomedical Analysis* 180: 113035.

Varki, A. 2017. Biological roles of glycans. *Glycobiology* 27: 3–49.

Veillon, L., C. Fakih, H. Abou-El-Hassan, F. Kobeissy, Y. Mechref. 2018. Glycosylation changes in brain cancer. *ACS Chemical Neuroscience* 9: 51–72.

Velkova, L., P. Dolashka, J. V. Beeumen, B. Devreese. 2017. N-glycan structures of beta-HlH subunit of helix lucorum hemocyanin. *Carbohydrate Research* 449: 1–10.

Wada, Y. 2016. Mass spectrometry of transferrin glycoforms to detect congenital disorders of glycosylation: Site-specific profiles and pitfalls. *Proteomics* 16: 3105–110.

Walsh, G. 2014. Biopharmaceutical benchmarks 2014. *Nature Biotechnology* 32(10): 992–1000.

Wang, M., X. M. Zhang, C. H. Deng. 2015. Facile synthesis of magnetic poly(styrene-co-4-vinylbenzene-boronic acid) microspheres for selective enrichment of glycopeptides. *Proteomics* 15: 2158–65.

Wang, J. X., R. Cunningham, H. Zetterberg, et al. 2016a. Label-free quantitative comparison of cerebrospinal fluid glycoproteins and endogenous peptides in subjects with Alzheimer's disease, mild cognitive impairment, and healthy individuals. *Proteomics: Clinical Applications* 10: 1225–41.

Wang, Z. H., C. Lorin, M. Koutsoukos, et al. 2016b. Comprehensive characterization of reference standard lots of HIV-1 subtype C Gp120 proteins for clinical trials in Southern African regions. *Vaccines* 4: 17.

Wang, J. R., W. N. Gao, R. Grimm, et al. 2017. A method to identify trace sulfated IgG N-glycans as biomarkers for rheumatoid arthritis. *Nature Communications* 8: 631.

Wang, X. N., Z. A. Deng, C. C. Huang, et al. 2018. Differential N-glycan patterns identified in lung adenocarcinoma by N-glycan profiling of formalin-fixed paraffin-embedded (FFPE) tissue sections. *Journal of Proteomics* 172: 1–10.

Wang, Y. Y., Y. Cai, Y. Zhang, H. J. Lu. 2019. Glycan reductive amino acid coded affinity tagging (GRACAT) for highly specific analysis of N-glycome by mass spectrometry. *Analytica Chimica Acta* 1089: 90–9.

Wang, Z. Y., J. W. Zhu, H. L. Lu. 2020a. Antibody glycosylation: Impact on antibody drug characteristics and quality control. *Applied Microbiology and Biotechnology* 104: 1905–14.

Wang, L. M., L. J. Yang, Y. Zhang, H. J. Lu. 2020b. Dual isotopic labeling combined with fluorous solid-phase extraction for simultaneous discovery of neutral/sialylated N-glycans as biomarkers for gastric cancer. *Analytica Chimica Acta* 1104: 87–94.

Wang, Y., Z. X. Tian. 2020. New energy setup strategy for intact N-glycopeptides characterization using higher-energy collisional dissociation. *Journal of the American Society for Mass Spectrometry* 31: 651–7.

Watanabe, Y., J. D. Allen, D. Wrapp, J. S. McLellan, M. Crispin. 2020. Site-specific glycan analysis of the SARS-CoV-2 spike. *Science* 369(6501): 330–3.

West, C. A., M. J. Wang, H. Herrera, et al. 2018. N-linked glycan branching and fucosylation are increased directly in HCC tissue as determined through *in situ* glycan imaging. *Journal of Proteome Research* 17: 3454–62.

Wu, R. Q., Y. Xie, C. H. Deng. 2016a. Thiol-ene click synthesis of L-cysteine-bonded zwitterionic hydrophilic magnetic nanoparticles for selective and efficient enrichment of glycopeptides. *Talanta* 160: 461–9.

Wu, R. Q., L. T. Li, C. H. Deng.. 2016b. Highly efficient and selective enrichment of glycopeptides using easily synthesized magG/PDA/Au/L-Cys composites. *Proteomics* 16: 1311–20.

Wu, G., P. G. Hitchen, M. Panico, et al. 2016c. Glycoproteomic studies of IgE from a novel hyper IgE syndrome linked to PGM3 mutation. *Glycoconjugate Journal* 33(3): 447–56.

Wu, D., W. B. Struwe, D. J. Harvey, M. A. J. Ferguson, C. V. Robinson. 2018. N-glycan microheterogeneity regulates interactions of plasma proteins. *Proceedings of the National Academy of Sciences USA* 115(35): 8763–8.

Wu, A. Y., K. Ueda, C. P. Lai.. 2019. Proteomic analysis of extracellular vesicles for cancer diagnostics. *Proteomics* 19: e1800162.

Wu, Y. L., N. R. Sun, C. H. Deng. 2020. Construction of magnetic covalent organic frameworks with inherent hydrophilicity for efficiently enriching endogenous glycopeptides in human saliva. *ACS Applied Materials & Interfaces* 12: 9814–23.

Wuhrer, M., M. H. Selman, L. A. McDonnell, et al. 2015. Pro-inflammatory pattern of IgG1 Fc glycosylation in multiple sclerosis cerebrospinal fluid. *Journal of Neuroinflammation* 12: 235.

Xie, Y. Q., Q. J. Liu, Y. Li, C. H. Deng. 2018. Core-shell structured magnetic metal-organic framework composites for highly selective detection of N-glycopeptides based on boronic acid affinity chromatography. *Journal of Chromatography A* 1540: 87–93.

Yang, X. Y., S. M. Kim, R. Ruzanski, et al. 2016a. Ultrafast and high-throughput N-glycan analysis for monoclonal antibodies. *MAbs* 8(4): 706–17.

Yang, Y., F. Liu, V. Franc, L. A. Halim, H. Schellekens, A. J. R. Heck. 2016b. Hybrid mass spectrometry approaches in glycoprotein analysis and their usage in scoring biosimilarity. *Nature Communications* 7: 13397.

Yang, S., L. J. Chen, D. W. Chan, Q. K. Li, H. Zhang. 2017. Protein signatures of molecular pathways in non-small cell lung carcinoma (NSCLC): comparison of glycoproteomics and global proteomics. *Clinical Proteomics* 14: 31.

Yang, L. J., X. X. Du, Y. Peng, et al. 2019. Integrated Pipeline of Isotopic Labeling and Selective Enriching for Quantitative Analysis of N-Glycome by Mass Spectrometry. *Analytical Chemistry* 91: 1486–93.

Yang, Y., Y. Du, I. A. Kaltashov. 2020. The utility of native MS for understanding the mechanism of action of repurposed therapeutics in COVID-19: heparin as a disruptor of the SARS-CoV-2 interaction with its host cell receptor. *Analytical Chemistry* 92(16):10930–4.

Yin, H. D., Z. J. Tan, J. Wu, et al. 2015. Mass-selected site-specific core-fucosylation of serum proteins in hepatocellular carcinoma. *Journal of Proteome Research* 14: 4876–84.

Yu, X. W., Y. X. Wang, J. Kristic, et al. 2016. Profiling IgG N-glycans as potential biomarker of chronological and biological ages. *Medicine* 95: 28.

Yuan, S. S., Q. Q. Li, Y. Zhang, et al. 2015a. Changes in anti-thyroglobulin IgG glycosylation patterns in Hashimoto's thyroiditis patients. *The Journal of Clinical Endocrinology and Metabolism* 100(2): 717–24.

Yuan, W., M. Sanda, J. Wu, J. Koomen, R. Goldman. 2015b. Quantitative analysis of immunoglobulin subclasses and subclass specific glycosylation by LC-MS-MRM in liver disease. *Journal of Proteomics* 116: 24–33.

Ząbczyńska, M., P. Link-Lenczowski, M. Novokmet, et al. 2020. Altered N-glycan profile of IgG-depleted serum proteins in Hashimoto's thyroiditis. *BBA - General Subjects* 1864: 129464.

Zhang, Y. W., J. H. Zhu, H. D. Yin, J. Marrero, X. X. Zhang, D. M. Lubman. 2015. ESI-LC-MS method for haptoglobin fucosylation analysis in hepatocellular carcinoma and liver cirrhosis. *Journal of Proteome Research* 14: 5388–95.

Zhang, D., B. C. Chen, Y. M. Wang, et al. 2016. Disease-specific IgG fc N-glycosylation as personalized biomarkers to differentiate gastric cancer from benign gastric diseases. *Scientific Reports* 6: 25957.

Zhang, Q., Z. Li, Y. Wang, Q. Zheng, J. Li. 2018. Mass spectrometry for protein sialoglycosylation. *Mass Spectrometry Reviews* 37: 652–80.

Zhang, Z., B. Shah, J. Richardson. 2019a. Jason. Impact of Fc N-glycan sialylation on IgG structure. *MAbs* 11(8): 1381–90.

Zhang, Y., H. Y. Jing, T. Wen, et al. 2019b. Phenylboronic acid functionalized C3N4 facultative hydrophilic materials for enhanced enrichment of glyco-peptides. *Talanta* 191: 509–18.

Zhang, Z. J., M. Westhrin, A. Bondt, M. Wuhrer, T. Standal, S. Holst. 2019c. Serum protein N-glycosylation changes in multiple myeloma. *Biochimica et Biophysica Acta - General Subjects* 1863: 960–70.

Zhang, Y., H. Y. Jing, B. Meng, X. Y. Qian, W. T. Ying. 2020a. L-cysteine func-tionalized straticulate C3N4 for the selective enrichment of glycopeptides. *Journal of Chromatography A* 1610: 460545.

Zhang, Y., Y. H. Mao, W. J. Zhao, et al. 2020b. Glyco-CPLL: An integrated method for in-depth And comprehensive N-glycoproteome profiling of human plasma. *Journal of Proteome Research* 19: 655–66.

Zhang, H. Q., Y. Y. Lv, J. Du, et al. 2020c. A GSH functionalized magnetic ultra-thin 2D-MoS2 nanocomposite for HILIC-based enrichment of N-glycopeptides from urine exosome and serum proteins. *Analytica Chimica Acta* 1098: 181–9.

Zhao, M. Z., Y. W. Zhang, F. Yuan. 2015. Hydrazino-s-triazine based labelling reagents for highly sensitive glycan analysis via liquid chromatography-electrospray mass spectrometry. *Talanta* 144: 992–7.

Zhao, D. D., B. Sun, S. Y. Sun, et al. 2017. Characterization of human entero-virus71 virus-like particles used for vaccine antigens. *PLoS One* 12(7): e0181182.

Zhao, X. Y., Y. Huang, G. Ma, et al. 2020. Parallel on-target derivatization for mass calibration and rapid profiling of N-glycans by MALDI-TOF MS. *Analytical Chemistry* 92: 991–8.

Zheng, H. J., J. X. Jia, Z. Li, Q. Jia. 2020. Bifunctional magnetic supramolecular-organic framework: a nanoprobe for simultaneous enrichment of glycosyl-ated and phosphorylated peptides. *Analytical Chemistry* 92: 2680–9.

Zhou, J. Y., H. Y. Gao, W. C. Xie, Y. Li. 2020a. Bovine serum albumin affects N-glycoforms of murine IgG monoclonal antibody purified from hybrid-oma supernatants. *Applied Microbiology and Biotechnology* 104: 1583–94.

Zhou, D. P., X. X. Tian, R. B. Qi, C. Peng, W. Zhang. 2020b. Identification of 22 N-glycosites on spike glycoprotein of SARS-CoV-2 and accessible surface glycopeptide motifs: implications for vaccination and antibody therapeu-tics. *Glycobiology* doi:. doi: 10.1093/glycob/cwaa052.

Zhu, J. H., J. Wu, H. D. Yin, J. Marrero, D. M. Lubman. 2015. Mass spectrometric N-glycan analysis of haptoglobin from patient serum samples using a 96-well plate format. *Journal of Proteome Research* 14: 4932–9.

Zhu, H. B., C. Qiu, A. C. Ruth, D. A. Keire, H. P. Ye. 2017. A LC-MS All-in-one workflow for site-specific location, identification and quantification of N-/O-glycosylation in human chorionic gonadotropin drug products. *The AAPS Journal* 19: 846–55.

Zhu, J. H., Z. W. Chen, J. Zhang, et al. 2019. Differential quantitative determination of site-specific intact N-Glycopeptides in serum Haptoglobin between hepatocellular carcinoma and cirrhosis using LC-EThcD-MS/MS. *Journal of Proteome Research* 18(1): 359–71.

Zou, G. Z., J. D. Benktander, S. T. Gizaw, S. Gaunitz, M. V. Novotny. 2017. Comprehensive analytical approach toward glycomic characterization and profiling in urinary exosomes. *Analytical Chemistry* 89(10): 5364–72.

Zou, Y., J. X. Hu, J. Z. Jie, et al. 2020. Comprehensive analysis of human IgG Fc N-glycopeptides and construction of a screening model for colorectal cancer. *Journal of Proteomics* 213: 103616.

Index

Printed in the United States
by Baker & Taylor Publisher Services